T0134876

Human Interphase Chromosomes

Ivan Iourov • Svetlana Vorsanova • Yuri Yurov
Editors

Human Interphase Chromosomes

Biomedical Aspects

 Springer

Editors
Ivan Iourov
Mental Health Research Center
Russian Academy of Medical Sciences
Moscow, Russia

Svetlana Vorsanova
Veltischev Research and Clinical Institute
Pirogov Russian National Research
Medical University
Moscow, Russia

Yuri Yurov
Mental Health Research Center
Russian Academy of Medical Sciences
Moscow, Russia

ISBN 978-3-030-62534-4   ISBN 978-3-030-62532-0   (eBook)
https://doi.org/10.1007/978-3-030-62532-0

This Springer imprint is published by the registered company Springer Nature Switzerland AG
The registered company address is: Gewerbestrasse 11, 6330 Cham, Switzerland

We dedicate this work to our close relative and colleague, Ilia V Soloviev, who will never be forgotten. He was a talented young researcher and a pioneer of molecular cytogenetics, genome, and chromosome research. His prodigious work and original ideas have formed our current research directions.

Dr. Ilia V Soloviev

The book is also dedicated to Prof. Yuri B Yurov, who left us in 2017. It is hardly possible to describe Yuri's contribution to bioscience and molecular cytogenetics. For more details, please see Iourov IY, Vorsanova SG: Yuri B. Yurov (1951–2017). Molecular Cytogenetics 2018; 11:36

Prof. Yuri B Yurov

Preface

In 2013, *Human Interphase Chromosomes—Biomedical Aspects* edited by Yuri B. Yurov, Svetlana G. Vorsanova, Ivan Y. Iourov (Springer Science+Business Media, LLC 2013; Print ISBN: 978-1-4614-6557-7; Online ISBN: 978-1-4614-6558-4) was published. Taking into account the success of that publication, we have accepted the kind invitation of Springer for issuing the second edition of *Human Interphase Chromosomes—Biomedical Aspects*. Tragically, our co-editor Prof. Yuri B Yurov passed away in 2017. Nevertheless, his name has to be among the editors and authors of chapters inasmuch as his original ideas and prodigious work underlies the content of the book.

The study of human interphase chromosomes is important for understanding eukaryotic DNA expression and replication as the interphase represents the essential period of cellular life. Knowledge about the architectural organization of chromosomes inside the nuclear space is important for understanding genome functioning during the cell cycle. Moreover, human chromosomal variations require the use of molecular cytogenetic techniques for interphase chromosomal analysis, because the human organism has >200 cell types, the majority of which are in interphase. As we noted in the preface to the previous edition: interphase cytogenetics *"is often viewed as an esoteric discipline that is only concerns few specialists trying to implement single-cell approaches to genome biology and medicine. However, studying interphase chromosomes is relevant to numerous fields of life sciences including but not limited to molecular and cell biology, biomedicine, genetics (including medical genetics), neuroscience, evolution, oncology, and genomics."*

The main body of the book is composed of nine chapters. Chapter 1 (by Prof. Ivan Iourov et al.) is devoted to Human Interphase Cytogenomics, "the rebranded research area integrates data on chromosomes acquired by visualization, array/sequencing and bioinformatics assays for understanding the 3D genome, molecular/cellular pathways and phenome in health and disease." Chapter 2 (contributed by Prof. SV Razin and his colleagues) is a brilliant description of spatial genome behavior in interphase. The third chapter (by Drs. JW Oh and A Abyzov) acknowledges current trends in analysis of cell and nuclear genome by next-generation sequencing providing the state of the art in studying cellular genomes at the DNA sequence level. Chapter 4 (by Prof. YB Yurov et al.) is dedicated to interphase chromosomes of the human brain. The role of chromosomal variation in the normal and diseased human brain is discussed. Drs. JM Bridger and HA Foster have described cellular senescence in the genomic/chromosomal context in Chap. 5. Unclassified chromosome abnormalities and genome behavior in interphase are described by Prof. H. Heng and his colleagues in Chap. 6. Chapter 7 (by Prof. SG Vorsanova and her colleagues) is dedicated to the role of interphase fluorescence *in situ* hybridization in current biomedical research and molecular diagnosis. Prof. T. Liehr has described the analysis of chromosome architecture using high-resolution FISH-banding in three-dimensionally preserved human interphase nuclei in Chap. 8. The final chapter (Chap. 9) expresses a chromosome-centric view on the genome. In the authors' opinion, *"there is an urgent need for expressing chromosome-centric concepts for filling the "chromosomal gap" in human genetics (genomics) and genomic medicine. To succeed, one has to look at the problem from different perspectives: theoretical, empirical, diagnostic, and educational."* To this end, we hope that the second edition of *Human Interphase Chromosomes—Biomedical Aspects* is able to repeat the success of the first edition.

Moscow, Russia   Ivan Y. Iourov
                                                                            Svetlana G. Vorsanova

Contents

Contributors

Batoul Abdallah Center for Molecular Medicine and Genomics, Wayne State University School of Medicine, Detroit, MI, USA

Alexej Abyzov Department of Health Sciences Research, Center for Individualized Medicine, Mayo Clinic, Rochester, MN, USA

Joanna M. Bridger Centre for Genome Engineering and Maintenance, Department of Life Sciences, Division of Biosciences, College of Health, Medicine and Life Sciences, Brunel University London, Uxbridge, UK

Helen A. Foster Department of Clinical, Pharmaceutical and Biological Science, School of Life and Medical Sciences, University of Hertfordshire, Hatfield, UK

Alexey A. Gavrilov Institute of Gene Biology, Russian Academy of Sciences, Moscow, Russia

Henry H. Heng Center for Molecular Medicine and Genomics, Wayne State University School of Medicine, Detroit, MI, USA

Department of Pathology, Wayne State University School of Medicine, Detroit, MI, USA

Steve Horne Center for Molecular Medicine and Genomics, Wayne State University School of Medicine, Detroit, MI, USA

Ivan Y. Iourov Veltischev Research and Clinical Institute for Pediatrics of the Pirogov Russian National Research Medical University, Ministry of Health of Russian Federation, Moscow, Russia

Mental Health Research Center, Moscow, Russia

Medical Genetics Department of Russian Medical Academy of Continuous Postgraduate Education, Moscow, Russia

Alexei D. Kolotii Veltischev Research and Clinical Institute for Pediatrics, Pirogov Russian National Research Medical University, Ministry of Health of Russian Federation, Moscow, Russia

Mental Health Research Center, Russian Ministry of Science and Higher Education, Moscow, Russia

Oxana S. Kurinnaia Veltischev Research and Clinical Institute for Pediatrics, Pirogov Russian National Research Medical University, Ministry of Health of Russian Federation, Moscow, Russia

Mental Health Research Center, Russian Ministry of Science and Higher Education, Moscow, Russia

Thomas Liehr Jena University Hospital, Friedrich Schiller University, Institute of Human Genetics, Jena, Germany

Guo Liu Center for Molecular Medicine and Genomics, Wayne State University School of Medicine, Detroit, MI, USA

Ji Won Oh Department of Anatomy, School of Medicine, Kyungpook National University, Daegu, South Korea

Biomedical Research Institute, Kyungpook National University Hospital, Daegu, South Korea

Sergey V. Razin Institute of Gene Biology, Russian Academy of Sciences, Moscow, Russia

Faculty of Biology, M.V. Lomonosov Moscow State University, Moscow, Russia

Sarah Regan Center for Molecular Medicine and Genomics, Wayne State University School of Medicine, Detroit, MI, USA

Sergey V. Ulianov Institute of Gene Biology, Russian Academy of Sciences, Moscow, Russia

Faculty of Biology, M.V. Lomonosov Moscow State University, Moscow, Russia

Svetlana G. Vorsanova Mental Health Research Center, Moscow, Russia

Veltischev Research and Clinical Institute for Pediatrics of the Pirogov Russian National Research Medical University, Ministry of Health of Russian Federation, Moscow, Russia

Christine J. Ye The Division of Hematology/Oncology, Department of Internal Medicine, University of Michigan, Ann Arbor, MI, USA

Yuri B. Yurov Mental Health Research Center, Moscow, Russia

Veltischev Research and Clinical Institute for Pediatrics, Pirogov Russian National Research Medical University, Ministry of Health of Russian Federation, Moscow, **Russia**

Chapter 1
Human Interphase Cytogenomics

Ivan Y. Iourov, Svetlana G. Vorsanova, and Yuri B. Yurov

Abstract Landmark discoveries in chromosome biology are intimately associated with introducing novel molecular technologies. Cytogenetic analysis remains the gold standard for technological advances in human genetics. However, since the resolution of the analysis is rather low (~5 Mb), numerous molecular technologies with a higher resolution have been introduced to cytogenetics. Among these, there is interphase fluorescence in situ hybridization, which has also become a "must-use" platform for studying human chromosomes in interphase. Subsequently, techniques for analyzing spatial chromatin organization (C techniques) and whole genomes at cellular level (single-cell array and sequencing techniques) have been developed. Although these methods have become technological breakthroughs, numerous structural and functional aspects of chromosomal organization in interphase remain to be elucidated. Here, the role of interphase chromosomal analysis in contemporary biomedicine is assessed. It is generally accepted that nuclear chromosome organization contributes to almost all key intranuclear processes in health and disease. Additionally, interphase chromosomal analysis sheds light on intercellular and interindividual genome variability. Acknowledging the trend in molecular cytogenetics initiated more than a decade ago, we have rebranded human interphase cytogenetics. Accordingly, the term has been changed to human interphase cytogenomics. The rebranded research area integrates data on chromosomes acquired by visualization, array/sequencing, and bioinformatics assays for understanding the 3D genome, molecular/cellular pathways, and phenome in health and disease.

I. Y. Iourov (✉)
Mental Health Research Center, Moscow, Russia

Veltischev Research and Clinical Institute for Pediatrics of the Pirogov Russian National Research Medical University, Ministry of Health of Russian Federation, Moscow, Russia

Medical Genetics Department of Russian Medical Academy of Continuous Postgraduate Education, Moscow, Russia

S. G. Vorsanova · Y. B. Yurov
Mental Health Research Center, Moscow, Russia

Veltischev Research and Clinical Institute for Pediatrics of the Pirogov Russian National Research Medical University, Ministry of Health of Russian Federation, Moscow, Russia

© Springer Nature Switzerland AG 2020
I. Iourov et al. (eds.), *Human Interphase Chromosomes*,
https://doi.org/10.1007/978-3-030-62532-0_1

1

Almost 30 years ago, it was postulated that studying chromosome organization in the interphase nucleus is unavoidable for proper understanding of molecular processes involving nucleic acids in a cell (Manuelidis 1990). Technological developments in cytogenetics (molecular cytogenetics) have always been the driving force of chromosome research (Ferguson-Smith 2015; Liehr 2017). On the long and winding road to chromosomal analysis at any stage of the cell cycle, the introduction of interphase fluorescence in situ hybridization (FISH) has been the long-awaited starting point for the real studies of chromosome structures beyond the metaphase chromosomes. These aspects of human interphase chromosomes have been reviewed historically in a chapter of the previous edition of this book (Yurov et al. 2013b). Regardless of the availability of numerous molecular approaches toward single-cell DNA analysis, interphase FISH-based methods apparently remain the essential technological strategies for unraveling spatial arrangement and structural behavior of whole chromosomes in eukaryotes (Vorsanova et al. 2010a; Liehr 2017; Iourov et al. 2019c; Hu et al. 2020). Nonetheless, taking into account the data, which may be provided by the methodological arsenal of current bioscience, it seems absurd to disintegrate genomic, cytogenetic, epigenetic, proteomic, metabolomic, biochemical, etc., knowledge. Accordingly, microarray (molecular karyotyping), sequencing, and chromatin analysis (i.e., chromosome conformation capture or 3C techniques) together with data acquired by molecular cytogenetic and banding cytogenetic analyses might be extremely useful for chromosome biology. The integrated knowledge would certainly form a blueprint of cellular homeostasis. In this instance, cytogenomics (i.e., cytogenetic studies in the genomic context as initially defined in Iourov et al. 2008) seems to become more important biomedical area than previously recognized.

Chromosomal order in the interphase nucleus has been a focus of biological research from the second half of the nineteenth century to the present (Rabl 1885; Cremer et al. 2020). During the second half of the twentieth century, several compatible models were proposed to describe chromosome/chromatin order in the interphase nucleus (Comings 1968, 1980; Vogel and Schroeder 1974; Manuelidis 1990). This line of research had been purely theoretic until in situ DNA hybridization (Pinkel et al. 1986; Vorsanova et al. 1986) allowed the direct analysis of interphase chromosomal structures in humans (overviewed in Yurov et al. 2013a and Liehr 2017). Chromosomal analysis in interphase has provided evidences that chromosomes are spatially arranged in the nucleus occupying "chromosome territories" to modulate genome behavior at the supramolecular (intranuclear) level (Cremer and Cremer 2010; Rouquette et al. 2010). Structurally and functionally, spatial chromosome arrangement in interphase nuclei correlates with genome organization at sequence and banding levels (Foster and Bridger 2005; Jabbari and Bernardi 2017). Therefore, it is not surprising that further studies have discovered the involvement of spatial interphase chromosome arrangement in such critical biological processes/phenomena as transcriptional regulation, DNA replication and reparation, genomic imprinting, genome stability maintenance, programmed cell death, development, aging, and evolution (see reviews by Bickmore and van Steensel 2013; Gasser 2016; Finn and Misteli 2019; Fritz et al. 2016, 2019; Seeber et al. 2018; Henry et al. 2019;

Ravi et al. 2020). Disease phenotypes are occasionally associated with specific nuclear chromosome architecture as well (Foster and Bridger 2005; Iourov 2012; Kemeny et al. 2018; Finn and Misteli 2019). Furthermore, changes in chromatin behavior and state are systematically associated with the spatial arrangement of interphase chromosomes (Rosa and Everaers 2008; Zhang and Wolynes 2015; Yu and Ren 2017; Chicano and Daban 2019). In total, one may conclude that the ubiquitous concept "Form Follows Functions" (3F) comprehensively describing the global functional three-dimensional (3D) organization of human genome (i.e., the first dimension, DNA sequence; the second dimension, chromatin; the third dimension, chromosomes) culminates in the spatial chromosome arrangement in interphase.

The genomic 3F-3D interplay has been further delineated by 3C-based techniques, which have been found useful for uncovering missing links between genome behavior and DNA arrangement in interphase nuclei (reviewed by Dekker and Mirny 2016; Han et al. 2018; Kempfer and Pombo 2020). These studies have formed the firm basis for a new concept of 3D genomics, which integrates data on chromatin organization and its impact on the genome behavior mediated by spatial DNA arrangement in interphase (Dekker and Mirny 2016; Yu and Ren 2017; Spielmann et al. 2018). 3D genomics' concept has also been found applicable to determine mechanisms for a variety of diseases (Chakraborty and Ay 2019). For instance, 3D genomic concepts have long been proposed as a basis of new paradigm or as a new frontier in brain diseases (Mitchell et al. 2014). However, one has to note that chromosomal organization in interphase nuclei of the human brain remains a kind of dark matter of neuroscience/biomedicine (Yurov et al. 2018). Notwithstanding, changes in 3D genome organization producing pathological cell phenotypes are clearly demonstrated in various cancers, and it has been suggested that action of 3D genome-disrupting drugs might be effective in anticancer therapy (Kantidze et al. 2020). Additionally, several complex and monogenic diseases seem to exhibit specific chromatin arrangement referred to 3D genome alterations (Chakraborty and Ay 2019). In summary, to advance the 3D genome concept and to understand the relevance of genomic 3F-3D interplay in health and disease, human interphase cytogenetic analysis appears to be required.

Interphase molecular cytogenetics encompasses an important set of methods for uncovering genomic variations. Until recently, chromosomal analysis in interphase has been almost exclusively based on FISH, which is used for detection of chromosome-specific DNAs (i.e., DNA located at pericentromeric heterochromatin or euchromatic regions) and, more rarely, whole chromosomes (Yurov et al. 1996; Ried 1998; Fung et al. 2000; Iourov et al. 2006b, 2019c; Arendt et al. 2009; Vorsanova et al. 2010a; Wang et al. 2016). Even nowadays, the application of interphase FISH may be almost as effective as whole-genome single-cell analysis (single-cell whole-genome sequencing) for studying aneuploidy and specific chromosomal rearrangements in interphase nuclei (Bakker et al. 2015; Yurov et al. 2018; Andriani et al. 2019). Moreover, FISH-based methods (e.g., interphase chromosome-specific multicolor banding) are able to detect interphase chromosome breaks and abnormal chromosomal behavior in interphase, which hallmark a variety of pathogenic processes and are undetectable by other methods (Iourov et al. 2006a, 2007, 2009a). Chromosomal DNA replication appears to be another phenomenon specifically requiring interphase FISH for the analysis (Vorsanova et al. 2001). Currently, there are two alternative platforms for detecting genomic/chromosomal changes in individual cells: single-cell whole-genome or targeted sequencing (or, more rarely, microarray analysis) (Wang et al. 2013; Gawad et al. 2016; Paolillo et al. 2019) and interphase FISH (Yurov et al. 1996; Fung et al. 2000; Iourov et al. 2006a, 2007, 2012, 2019c; Andriani et al. 2019). Thus, interphase FISH-based techniques remain an important technological part of studying genomic variations despite of developments in single-cell analyses of DNA fractions by sequencing and microarray.

The importance of uncovering genomic variations in single cells of tissues, which are inappropriate for cytogenetic methods, has been consistently recognized. Interphase FISH has been found applicable for analyzing aneuploidy/polyploidy during prenatal development (Vorsanova et al. 2005; Yurov et al. 2005, 2007a; Russo et al. 2016). To achieve high efficiency in postnatal diagnosis of chromosomal mosaicism, interphase FISH is performed as well (Vorsanova et al. 2010b; Jackson-Cook 2011; Iourov et al. 2019c). More importantly, molecular cytogenetic analyses in brain diseases using interphase FISH-based approaches have allowed the discovery of new mechanisms for psychiatric and neurodegenerative diseases (Yurov et al. 2001, 2007b, 2018, 2019; Arendt et al. 2009; Iourov et al. 2009b, 2013; Frade and Gage 2017; Graham et al. 2019). Chromosomal instability and specific chromosomal rearrangements detectable in almost all types of cancers are repeatedly addressed by interphase FISH (Ried 1998; Nordgren et al. 2002; Liehr 2017). Actually, data acquired by interphase FISH has contributed to our understanding of aneuploidy's role and system (fuzzy) inheritance in such a devastating condition as cancer (Christine et al. 2018). Finally, human aging mediated by the accumulation of somatic chromosomal mutations (e.g., aneuploidy) is a focus of studies performed by interphase molecular cytogenetic techniques (Yurov et al. 2010; Zhang and Vijg 2018). To put the molecular cytogenetic data into the genomic context, one can suggest to integrate data acquired by visualization techniques (mainly, interphase FISH) and techniques for DNA fraction analysis (i.e., array/sequencing). For succeeding in the integration followed by the interpretation of data on genome variations, bioinformatic approaches/systems analyses are to be used. Our previous in silico molecular cytogenetic analyses, for example, have demonstrated that systems analysis of functional consequences of chromosomal imbalances and copy number variations may be useful for linking intercellular and interindividual genome variability (Iourov et al. 2014, 2019b; Iourov 2019a). Since the methodology encompassing visualization techniques, methods for studying DNA fractions (single-cell/multiple-cell) and bioinformatics (cytogenomics) may determine functioning of complex genetic systems at sequence, chromatin, and chromosome levels, one can suggest the success of introducing cytogenomic paradigm to interphase cytogenetics.

Cytogenomics is the study of chromosomes in the genomic context or inversely the study of the genome (genome variability) in chromosomal context (Iourov et al.

2008). Defined this way, cytogenomics has gained significant momentum in recent years (Bernheim 2010; Silva et al. 2019). Postgenomics methodology has further significantly contributed to cytogenomics' development (cytopostgenomics), opening new opportunities through systems biology (medicine) analysis (Iourov 2019b). In the postgenomic perspective, chromosome research requires massive data sets of genomic variations, transcriptome, proteome (interactome), and metabolome (Heng et al. 2018; Iourov 2019a, b). In terms of the chromosomal variation (variability of spatial chromosome arrangement + genomic variations at the chromosomal or sub-chromosomal level) in single cells, genomics (cytogenomics) using system biology has already been found useful to identify causes and consequences of genomic variations in individual cells (Iourov et al. 2012; Wang et al. 2013; Paolillo et al. 2019). Similarly, single-cell transcriptomic analyses have been successfully used for related purposes (Stubbington et al. 2017). The next task for the 3D genomics research would be the integration of (cyto)genomic and epigenomic data with the results of studying chromatin behavior. Fortunately, there have been developed a panel of protocols to succeed in empirical and in silico chromatin analysis (Woodcock and Ghosh 2010; Ramani et al. 2016; Weinreb and Raphael 2016). The description of the 3D genome requires, thereby, the completed sets of cytogenomic data.

Chromosomal analysis in interphase nuclei has generally been referred to as interphase molecular cytogenetics (Yurov et al. 2013b). In rapidly evolving research fields, rebranding of terms appears to be impending. In our opinion, taking into account the aforementioned rationale of modern interphase chromosome analysis, molecular cytogenetic studies of human interphase chromosomes may be designated as human interphase cytogenomics.

How does human interphase cytogenomics work? What place does it have in current biomedicine? Summarizing the previous ideas, it works as follows. Firstly, cytogenomic data acquired by visualization techniques (cytogenetic, molecular cytogenetic, and/or cytochemical methods) and techniques for analysis of DNA fractions (microarray/sequencing) are integrated using bioinformatics (systems analysis of genome variations at genome, epigenome, proteome, and metabolome levels). The acquired knowledge underlies interphase cytogenomics as a biomedical discipline. Secondly, using high-resolution 3C-/4C-/5C-based techniques for chromatin analysis, interphase cytogenomics plus chromatin data provide data on the 3D genome. Thirdly, systems analysis, applied to molecular, cellular, and tissular pathways in the 3D genomic context, is an apparent basis for molecular bioscience. To achieve a (bio)medical relevance, this global knowledge should be correlated with the results of phenome analysis. Visually, the way how human interphase cytogenomics works and its place in current biomedicine may be schematically presented as shown by Fig. 1.1.

If cytogenomic data are classified using pathways, one may uncover the molecular and cellular processes altered/modified by the genomic variations (for more details, see Iourov et al. 2019b). Slowly but surely, current medicine moves toward a systems science. As a result, pathway-based classification has become an important element of almost all representative biomedical studies, suggesting a

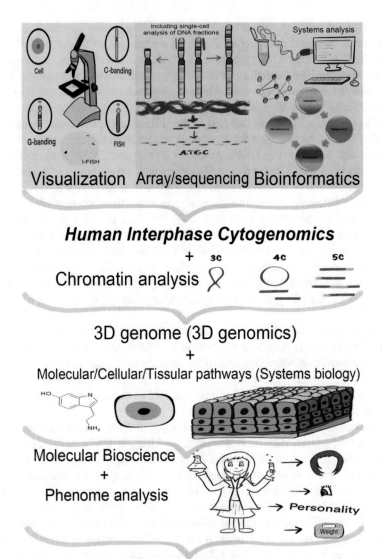

Fig. 1.1 Human interphase cytogenomics and its place in current biomedicine. The basis of human interphase cytogenomics is formed by knowledge acquired from visualization (micros-copy/(molecular) cytogenetic techniques; greenish area) and multiple-cell/single-cell of DNA fractions by array-/sequencing-based techniques (reddish area) integrated by systems analysis using in silico analyses of genome, epigenome, proteome, and metabolome (bioinformatics; bluish area). Interphase cytogenomic data with chromatin analysis (3C-/4C-/5C-based techniques) is the way to the reconstruction of the 3D genome (3D genomics). Systems biology analysis of molecu-lar, cellular, and tissular pathways in the 3D genomic context (molecular bioscience) correlated with detailed phenome analysis may be the essence of current biomedicine

reevaluation of disease etiology in an unprecedented manner (Iourov et al. 2019a). Unfortunately, neither the results of interphase chromosomal analyses nor the data on spatial chromosome organization have been addressed by a pathway-based (systems biology or cyto(post)genomic) analysis. Since studies dedicated to human interphase cytogenomics have the potential to complement our understanding of genome behavior at the supramolecular/chromosomal level at all stages of cell cycle in health and disease, it is important to incorporate cytogenomic data on interphase chromosomes into global omics' data sets.

Acknowledgments We express our gratitude to Dr. Maria A. Zelenova for Fig. 1.1. Professors SG Vorsanova and IY Iourov are partially supported by RFBR and CITMA according to the research project No. 18-515-34005. Prof. IY Iourov's lab is supported by the Government Assignment of the Russian Ministry of Science and Higher Education, Assignment no. AAAA-A19-119040490101-6. Prof. SG Vorsanova's lab is supported by the Government Assignment of the Russian Ministry of Health, Assignment no. AAAA-A18-118051590122-7.

References

Andriani GA, Maggi E, Piqué D et al (2019) A direct comparison of interphase FISH versus low-coverage single cell sequencing to detect aneuploidy reveals respective strengths and weaknesses. Sci Rep 9(1):10508

Arendt T, Mosch B, Morawski M (2009) Neuronal aneuploidy in health and disease: a cytomic approach to understand the molecular individuality of neurons. Int J Mol Sci 10(4):1609–1627

Bakker B, van den Bos H, Lansdorp PM et al (2015) How to count chromosomes in a cell: an overview of current and novel technologies. BioEssays 37(5):570–577

Bernheim A (2010) Cytogenomics of cancers: from chromosome to sequence. Mol Oncol 4(4):309–322

Bickmore WA, van Steensel B (2013) Genome architecture: domain organization of interphase chromosomes. Cell 152(6):1270–1284

Chakraborty A, Ay F (2019) The role of 3D genome organization in disease: from compartments to single nucleotides. Semin Cell Dev Biol 90:104–113

Chicano A, Daban JR (2019) Chromatin plates in the interphase nucleus. FEBS Lett 593(8):810–819

Christine JY, Regan S, Liu G et al (2018) Understanding aneuploidy in cancer through the lens of system inheritance, fuzzy inheritance and emergence of new genome systems. Mol Cytogenet 11:31

Comings DE (1968) The rationale for an ordered arrangement of chromatin in the interphase nucleus. Am J Hum Genet 20:440–460

Comings DE (1980) Arrangement of chromatin in the nucleus. Hum Genet 53(2):131–143

Cremer T, Cremer M (2010) Chromosome territories. Cold Spring Harb Perspect Biol 2(3):a003889

Cremer T, Cremer M, Hübner B et al (2020) The interchromatin compartment participates in the structural and functional organization of the cell nucleus. BioEssays 42(2):e1900132

Dekker J, Mirny L (2016) The 3D genome as moderator of chromosomal communication. Cell 164(6):1110–1121

Ferguson-Smith MA (2015) History and evolution of cytogenetics. Mol Cytogenet 8:19

Finn EH, Misteli T (2019) Molecular basis and biological function of variability in spatial genome organization. Science 365(6457):caaw9498

Foster HA, Bridger JM (2005) The genome and the nucleus: a marriage made by evolution. Genome organisation and nuclear architecture. Chromosoma 114(4):212–229

Frade JM, Gage FH (2017) Genomic mosaicism in neurons and other cell types. Springer, New York

Fritz AJ, Barutcu AR, Martin-Buley L et al (2016) Chromosomes at work: organization of chromosome territories in the interphase nucleus. J Cell Biochem 117(1):9–19

Fritz AJ, Sehgal N, Pliss A et al (2019) Chromosome territories and the global regulation of the genome. Genes Chromosomes Cancer 58(7):407–426

Fung J, Weier HU, Goldberg JD et al (2000) Multilocus genetic analysis of single interphase cells by spectral imaging. Hum Genet 107(6):615–622

Gasser SM (2016) Nuclear architecture: past and future tense. Trends Cell Biol 26(7):473–475

Gawad C, Koh W, Quake SR (2016) Single-cell genome sequencing: current state of the science. Nat Rev Genet 17(3):175–188

Graham EJ, Vermeulen M, Vardarajan B et al (2019) Somatic mosaicism of sex chromosomes in the blood and brain. Brain Res 1721:146345

Han J, Zhang Z, Wang K (2018) 3C and 3C-based techniques: the powerful tools for spatial genome organization deciphering. Mol Cytogenet 11:21

Heng HH, Horne SD, Chaudhry S et al (2018) A postgenomic perspective on molecular cytogenetics. Curr Genomics 19(3):227–239

Henry MP, Hawkins JR, Boyle J et al (2019) The genomic health of human pluripotent stem cells: genomic instability and the consequences on nuclear organization. Front Genet 9:623

Hu Q, Maurais EG, Ly P (2020) Cellular and genomic approaches for exploring structural chromosomal rearrangements. Chromosom Res 28:19–30

Iourov IY (2012) To see an interphase chromosome or: how a disease can be associated with specific nuclear genome organization. BioDiscovery 4:e8932

Iourov IY (2019a) Cytogenomic bioinformatics: practical issues. Cur Bioinformatics 14(5):372–373

Iourov IY (2019b) Cytopostgenomics: what is it and how does it work? Curr Genomics 20(2):77–78

Iourov IY, Liehr T, Vorsanova SG et al (2006a) Visualization of interphase chromosomes in postmitotic cells of the human brain by multicolour banding (MCB). Chromosom Res 14(3):223–229

Iourov IY, Vorsanova SG, Yurov YB (2006b) Chromosomal variation in mammalian neuronal cells: known facts and attractive hypotheses. Int Rev Cytol 249:143–191

Iourov IY, Liehr T, Vorsanova SG et al (2007) Interphase chromosome-specific multicolor banding (ICS-MCB): a new tool for analysis of interphase chromosomes in their integrity. Biomol Eng 24(4):415–417

Iourov IY, Vorsanova SG, Yurov YB (2008) Molecular cytogenetics and cytogenomics of brain diseases. Curr Genomics 9(7):452–465

Iourov IY, Vorsanova SG, Liehr T et al (2009a) Increased chromosome instability dramatically disrupts neural genome integrity and mediates cerebellar degeneration in the ataxia-telangiectasia brain. Hum Mol Genet 18(14):2656–2669

Iourov IY, Vorsanova SG, Liehr T et al (2009b) Aneuploidy in the normal, Alzheimer's disease and ataxia-telangiectasia brain: differential expression and pathological meaning. Neurobiol Dis 34(2):212–220

Iourov IY, Vorsanova SG, Yurov YB (2012) Single cell genomics of the brain: focus on neuronal diversity and neuropsychiatric diseases. Curr Genomics 13(6):477–488

Iourov IY, Vorsanova SG, Yurov YB (2013) Somatic cell genomics of brain disorders: a new opportunity to clarify genetic-environmental interactions. Cytogenet Genome Res 139(3):181–188

Iourov IY, Vorsanova SG, Yurov YB (2014) In silico molecular cytogenetics: a bioinformatic approach to prioritization of candidate genes and copy number variations for basic and clinical genome research. Mol Cytogenet 7(1):98

Iourov IY, Vorsanova SG, Yurov YB (2019a) Pathway-based classification of genetic diseases. Mol Cytogenet 12:4

Iourov IY, Vorsanova SG, Yurov YB (2019b) The variome concept: focus on CNVariome. Mol Cytogenet 12:52

Iourov IY, Vorsanova SG, Yurov YB et al (2019c) Ontogenetic and pathogenetic views on somatic chromosomal mosaicism. Genes (Basel) 10(5):E379

Jabbari K, Bernardi G (2017) An isochore framework underlies chromatin architecture. PLoS One 12(1):e0168023

Jackson-Cook C (2011) Constitutional and acquired autosomal aneuploidy. Clin Lab Med 31(4):481–511

Kantidze OL, Gurova KV, Studitsky VM et al (2020) The 3D genome as a target for anticancer therapy. Trends Mol Med 26(2):141–149

Kemeny S, Tatout C, Salaun G et al (2018) Spatial organization of chromosome territories in the interphase nucleus of trisomy 21 cells. Chromosoma 127(2):247–259

Kempfer R, Pombo A (2020) Methods for mapping 3D chromosome architecture. Nat Rev Genet 21(4):207–226

Liehr T (2017) Fluorescence *in situ* hybridization (FISH) — application guide. Springer, Berlin Heidelberg

Manuelidis L (1990) A view of interphase chromosomes. Science 250(4987):1533–1540

Mitchell AC, Bharadwaj R, Whittle C et al (2014) The genome in three dimensions: a new frontier in human brain research. Biol Psychiatry 75(12):961–969

Nordgren A, Heyman M, Sahlén S et al (2002) Spectral karyotyping and interphase FISH reveal abnormalities not detected by conventional G-banding: implications for treatment stratification of childhood acute lymphoblastic leukaemia: detailed analysis of 70 cases. Eur J Haematol 68(1):31–41

Paolillo C, Londin E, Fortina P (2019) Single-cell genomics. Clin Chem 65(8):972–985

Pinkel D, Straume T, Gray JW (1986) Cytogenetic analysis using quantitative, high-sensitivity, fluorescence hybridization. Proc Natl Acad Sci U S A 83(9):2934–2938

Rabl C (1885) Uber Zelltheilung. In: Gegenbaur C (ed) Morphologisches Jahrbuch 10:214–330

Ramani V, Shendure J, Duan Z (2016) Understanding spatial genome organization: methods and insights. Genomics Proteomics Bioinformatics 14(1):7–20

Ravi M, Ramanathan S, Krishna K (2020) Factors, mechanisms and implications of chromatin condensation and chromosomal structural maintenance through the cell cycle. J Cell Physiol 235(2):758–775

Ried T (1998) Interphase cytogenetics and its role in molecular diagnostics of solid tumors. Am J Pathol 152(2):325–327

Rosa A, Everaers R (2008) Structure and dynamics of interphase chromosomes. PLoS Comput Biol 4(8):e1000153

Rouquette J, Cremer C, Cremer T et al (2010) Functional nuclear architecture studied by microscopy: present and future. Int Rev Cell Mol Biol 282:1–90

Russo R, Sessa AM, Fumo R et al (2016) Chromosomal anomalies in early spontaneous abortions: interphase FISH analysis on 855 FFPE first trimester abortions. Prenat Diagn 36(2):186–191

Seeber A, Hauer MH, Gasser SM (2018) Chromosome dynamics in response to DNA damage. Annu Rev Genet 52:295–319

Silva M, de Leeuw N, Mann K et al (2019) European guidelines for constitutional cytogenomic analysis. Eur J Hum Genet 27(1):1–16

Spielmann M, Lupiáñez DG, Mundlos S (2018) Structural variation in the 3D genome. Nat Rev Genet 19(7):453–467

Stubbington MJ, Rozenblatt-Rosen O, Regev A et al (2017) Single-cell transcriptomics to explore the immune system in health and disease. Science 358(6359):58–63

Vogel F, Schroeder TM (1974) The internal order of the interphase nucleus. Humangenetik 25(4):265–297

Vorsanova SG, Yurov YB, Alexandrov IA et al (1986) 18p- syndrome: an unusual case and diagnosis by *in situ* hybridization with chromosome 18-specific alphoid DNA sequence. Hum Genet 72:185–187

Vorsanova SG, Yurov YB, Kolotii AD et al (2001) FISH analysis of replication and transcription of chromosome X loci: new approach for genetic analysis of Rett syndrome. Brain Dev 23(Sup.1):S191–S195

Vorsanova SG, Kolotii AD, Iourov IY et al (2005) Evidence for high frequency of chromosomal mosaicism in spontaneous abortions revealed by interphase FISH analysis. J Histochem Cytochem 53(3):375–380

Vorsanova SG, Yurov YB, Iourov IY (2010a) Human interphase chromosomes: a review of available molecular cytogenetic technologies. Mol Cytogenet 3:1

Vorsanova SG, Yurov YB, Soloviev IV et al (2010b) Molecular cytogenetic diagnosis and somatic genome variations. Curr Genomics 11(6):440–446

Wang Q, Zhu X, Feng Y et al (2013) Single-cell genomics: an overview. Front Biol 8(6):569–576

Wang H, La Russa M, Qi LS (2016) CRISPR/Cas9 in genome editing and beyond. Annu Rev Biochem 85:227–264

Weinreb C, Raphael BJ (2016) Identification of hierarchical chromatin domains. Bioinformatics 32(11):1601–1609

Woodcock CL, Ghosh RP (2010) Chromatin higher-order structure and dynamics. Cold Spring Harb Perspect Biol 2(5):a000596

Yu M, Ren B (2017) The three-dimensional organization of mammalian genomes. Annu Rev Cell Dev Biol 33:265–289

Yurov YB, Soloviev IV, Vorsanova SG et al (1996) High resolution multicolor fluorescence *in situ* hybridization using cyanine and fluorescein dyes: rapid chromosome identification by directly fluorescently labeled alphoid DNA probes. Hum Genet 97(3):390–398

Yurov YB, Vostrikov VM, Vorsanova SG et al (2001) Multicolor fluorescent *in situ* hybridization on post-mortem brain in schizophrenia as an approach for identification of low-level chromosomal aneuploidy in neuropsychiatric diseases. Brain Dev 23(Sup.1):S186–S190

Yurov YB, Iourov IY, Monakhov VV et al (2005) The variation of aneuploidy frequency in the developing and adult human brain revealed by an interphase FISH study. J Histochem Cytochem 53(3):385–390

Yurov YB, Iourov IY, Vorsanova SG et al (2007a) Aneuploidy and confined chromosomal mosaicism in the developing human brain. PLoS One 2(6):e558

Yurov YB, Vorsanova SG, Iourov IY et al (2007b) Unexplained autism is frequently associated with low-level mosaic aneuploidy. J Med Genet 44(8):521–525

Yurov YB, Vorsanova SG, Iourov IY (2010) Ontogenetic variation of the human genome. Curr Genomics 11(6):420–425

Yurov YB, Vorsanova SG, Iourov IY (2013a) Human interphase chromosomes — biomedical aspects. Springer New York, Heidelberg, Dordrecht, London

Yurov YB, Vorsanova SG, Iourov IY (2013b) Introduction to interphase molecular cytogenetics. In: Yurov YB, Vorsanova SG, Iourov IY (eds) Human interphase chromosomes — biomedical aspects. Springer New York, Heidelberg, Dordrecht, London, pp 1–8

Yurov YB, Vorsanova SG, Iourov IY (2018) Human molecular neurocytogenetics. Curr Genet Med Rep 6(4):155–164

Yurov YB, Vorsanova SG, Iourov IY (2019) Chromosome instability in the neurodegenerating brain. Front Genet 10:892

Zhang L, Vijg J (2018) Somatic mutagenesis in mammals and its implications for human disease and aging. Annu Rev Genet 52:397–419

Zhang B, Wolynes PG (2015) Topology, structures, and energy landscapes of human chromosomes. Proc Natl Acad Sci U S A 112(19):6062–6067

Chapter 2
Eukaryotic Genome in Three Dimensions

Sergey V. Razin, Alexey A. Gavrilov, and Sergey V. Ulianov

Abstract Modern ideas regarding the three-dimensional organization of the genome and its role in controlling gene expression are largely based on the results of research performed using the proximity ligation protocol. It has been demonstrated that genome folding is much less regular than was previously assumed. On the other hand, the genome was found partitioned into semi-independent structural-functional units commonly referred to as topologically associating domains (TADs). TAD borders restrict the areas of enhancer action via interfering with establishment of long-distance enhancer-promoter contacts. Within TADs, spatial juxtaposing of promoters to various enhancers or silencers results in the assembly of activating or repressing chromatin hubs that constitute an important part of epigenetic mechanisms regulating gene expression in higher eukaryotes. Within the cell nucleus, the spatial organization of the genome is tightly connected with functional compartmentalization of the nucleus. Recent evidence suggests that liquid phase separation plays an important role in establishing both the 3D genome organization and nuclear compartmentalization. In this chapter, we review the present state and outline the most important trends for future research in the area of 3D genomics.

Introduction

Studies of the 3D genome organization have become a trend in modern genomics. One may say that modern genomics has acquired a third dimension. As is often the case in science, a new stage in the study of genome organization and functioning was predetermined by the development of appropriate research tools. One biochemical protocol that had a major impact on the development of 3D genomics is the chromosome conformation capture protocol (Dekker et al. 2002). The main steps of this protocol are presented in Fig. 2.1. The key step of this procedure is

S. V. Razin (✉) · S. V. Ulianov
Institute of Gene Biology, Russian Academy of Sciences, Moscow, Russia

Faculty of Biology, M. V. Lomonosov Moscow State University, Moscow, Russia

A. A. Gavrilov
Institute of Gene Biology, Russian Academy of Sciences, Moscow, Russia

© Springer Nature Switzerland AG 2020
I. Iourov et al. (eds.), *Human Interphase Chromosomes*,
https://doi.org/10.1007/978-3-030-62532-0_2

Fig. 2.1 Main steps of the chromosome conformation capture protocol. Restriction enzymes are used to cut chromatin in intact nuclei isolated from formaldehyde-fixed cells. DNA fragments located in close proximity to each other are ligated with the T4 DNA ligase. qPCR or next-generation sequencing are used for the analysis of DNA chimeras obtained

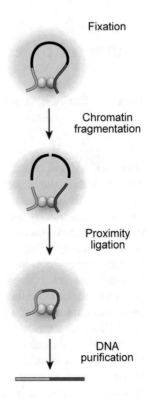

Fixation

Chromatin fragmentation

Proximity ligation

DNA purification

introduction of breaks into DNA within a fixed nucleus, followed by cross-ligation of closely located ends of broken DNA. Joining of DNA fragments located far from each other on the DNA chain but close in physical space creates chimeric DNA sequences containing information about the spatial proximity of the corresponding segments of genomic DNA. Analysis of the pools of chimeric fragments allows reconstructing the spatial organization of the genome based on the sets of captured pairwise interactions. This procedure was first successfully used to demonstrate that all remote enhancers of mouse beta-globin genes along with the promoters of genes, which are actually expressed, are organized into a common active chromatin hub (Tolhuis et al. 2002; de Laat and Grosveld 2003). This work highlighted the importance of 3D genome organization for the regulation of transcription. It has long been assumed that, to activate a gene, an enhancer should be in direct contact with this gene ((Bondarenko et al. 2003; West and Fraser 2005; Vernimmen and Bickmore 2015) and references herein). Taking into account that most enhancers are located far from the target gene, the ideal solution is to loop out the intervening segment of DNA, and 3C analysis has demonstrated that such situations are indeed quite common (Tolhuis et al. 2002; de Laat and Grosveld 2003; Gavrilov and Razin 2008; Philonenko et al. 2009; Vernimmen et al. 2007; Vernimmen et al. 2009). The number of enhancers in mammalian and Drosophila cells exceeds at least ten times the number of genes (Arnold et al. 2013; Consortium et al. 2012). The possibility of gene activation by different combinations of enhancers likely increases the regulatory capacity of the eukaryotic cell transcription control system. Disclosure of the functionally dependent mouse beta-globin gene domain 3D organization (Tolhuis et al. 2002; de Laat and Grosveld 2003) demonstrated for the first time how one gene or group of genes can be simultaneously activated by one or several enhancers.

The original 3C procedure allowed studying interactions between various regions within individual genomic loci. Eventually, various derivative procedures were developed collectively known as C-methods (reviewed in de Wit and de Laat (2012)). Most of these procedures, such as 4C (van de Werken et al. 2012), Hi-C (Lieberman-Aiden et al. 2009), and ChIA-PET (Fullwood et al. 2009), allowed performing genome-wide analysis. Application of these experimental protocols has provided deep insights into the role of 3D genome organization in transcription control (Denker and de Laat 2016; Dekker and Mirny 2016; Valton and Dekker 2016; Krijger and de Laat 2016). Of special importance, the genome was found to be partitioned into semi-independent self-interacting domains termed topologically associating domains or TADs (Nora et al. 2012; Dixon et al. 2012; Sexton et al. 2012). TADs appear to restrict the areas of enhancer action and thus can be considered as structural-functional units of the eukaryotic genome (Symmons et al. 2014, 2016). Disruption of TAD borders results in development of various genetic diseases (Lupianez et al. 2015, 2016; Krumm and Duan 2018; Franke et al. 2016). In normal situations, the patterns of enhancer-promoter spatial interactions change in the course of cell differentiation accordingly to activation and/or repression of particular genes. However, most of these changes occur within TADs while the TAD borders remain relatively stable (Dixon et al. 2016; Fraser et al. 2015). Nevertheless, a certain fraction of TAD boundaries is changed in the course of cell differentiation (Bonev and Cavalli 2016). To obtain further insights into mechanisms of eukaryotic genome functioning, it is highly important to disclose the nature of both TADs and TAD borders. This task is complicated by the fact that in virtually all eukaryotic cells studied, the contact chromatin domains are hierarchical (i.e., within larger domains, it is possible to annotate several levels of smaller and more dense nested domains) (Phillips-Cremins et al. 2013; Luzhin et al. 2019; Weinreb and Raphael 2016). It is not always obvious the domain of which level should be considered as TADs. Some authors claim that TADs can be discriminated only based on their functionality (i.e., as functional units of the genome rather than the units of a particular level of hierarchical genome folding) (Zhan et al. 2017). In this review, we shall discuss mechanisms of TAD formation and the impact of TADs on genome functioning.

Hierarchical Model of DNA Packaging in Chromatin

In most textbooks, it is possible to read that, in eukaryotic cells, genomic DNA is sequentially folded into 10 nm chromatin fiber (nucleosomal chain), into 30 nm chromatin fiber (which is frequently represented as a solenoid or zigzag), and then into loops of 30 nm fiber or several levels of "supersolenoid" structures (Fig. 2.2). Remarkably, this model of chromatin folding into regular structures was proposed approximately 30 years ago (Getzenberg et al. 1991) and is poorly supported by recent data. On the contrary, it is becoming increasingly evident that the only regular level of genomic DNA folding is wrapping of DNA around the octamers of nucleosome histones, resulting in formation of 10 nm fibers (Fussner et al. 2012). The latter then aggregate to form more or less compact chromatin masses (Maeshima et al. 2014a, b, 2016). Aggregation of chromatin fibers is promoted under conditions of macromolecular crowding (Hancock 2008) typical for nucleoplasm. Although at a medium scale thus formed chromatin masses appear irregular, at larger scales, they are subdivided into self-interacting domains that are commonly interpreted as chromatin globules. Such chromatin globules were observed in a high-resolution microscopic study of cell nuclei hybridized to chromosome- or locus-specific probes (Markaki et al. 2012; Smeets et al. 2014). Furthermore, the same structures

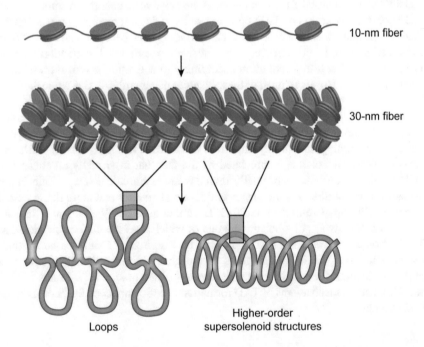

Fig. 2.2 A classical view of hierarchical folding of DNA in the nucleus. 10 nm nucleosome fiber folds into 30 nm fiber of variable architecture, which then forms hierarchical loops and supersolenoid structures

(1 Mb chromatin clusters) appear to correspond to early replicating chromatin domains (Markaki et al. 2010). In a recent study by the Cavalli laboratory, it was directly shown that TADs correspond to chromatin globules that can be visualized using FISH with TAD- and locus-specific probes (Szabo et al. 2018). Within the entire chromatin domain, TADs containing mostly active and mostly repressed chromatin are spatially segregated into the so-called A and B chromatin compartments, which likely correspond to euchromatin and heterochromatin (Lieberman-Aiden et al. 2009; Gibcus and Dekker 2013; Eagen 2018).

Most of the current knowledge about higher levels of DNA packaging in chromatin is based on the results of Hi-C analysis. The contact chromatin domains were observed in different taxa including mammals (Dixon et al. 2012; Nora et al. 2012), insects (Sexton et al. 2012), and birds (Ulianov et al. 2017). Of note, in Drosophila, TADs have a size in the range of 100 Kb (Sexton et al. 2012; Hou et al. 2012), while mammalian TADs are ten times larger (Dixon et al. 2012, 2016). Some contact domains can also be revealed in the genomes of plants and lower eukaryotes (Wang et al. 2015; Hsieh et al. 2015; Eser et al. 2017; Nikolaou 2017). However, they are substantially different from the TADs of mammals and Drosophila both in size and in the levels of insulation and genome coverage.

Interpretation of Hi-C maps strongly depends on resolution of the analysis. At 1 Mb resolution, only segregation of active and inactive chromatin can be registered (Lieberman-Aiden et al. 2009). 20–100 Kb resolution revealed TADs (Dixon et al. 2012, 2016; Gibcus and Dekker 2013). Finally, 1 Kb resolution maps demonstrated that TADs comprise two types of self-interacting domains, namely, looped domains and ordinary domains (Rao et al. 2014). The distinctive feature of looped domains in Hi-C maps is a spot at the top of a triangle reflecting a spatial proximity of loop bases (Fig. 2.3). In mammalian cells, chromatin loops originate due to enhancer-promoter interaction (Jin et al. 2013; Sahlen et al. 2015; Ghavi-Helm et al. 2014) or

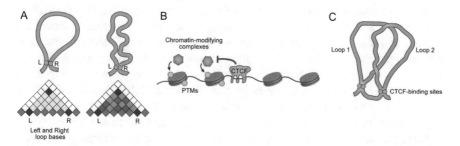

Fig. 2.3 Potential role of CTCF in defining chromatin spatial organization and epigenetic state. (**a**) Chromatin loop is manifested as a filled triangle in the Hi-C heat map only if numerous interactions between loop internal regions occur. (**b**) In a "traffic jam" model, DNA-bound CTCF restricts the spreading of histone posttranslation modifications along the chromatin fiber, preventing binding of chromatin-modifying complexes to nucleosomes located downstream of the CTCF-binding site. (**c**) Point-to-point interactions between CTCF-binding sites are unable to insulate extended loops from each other in the 3D nuclear space

because of interactions between CTCF-binding sites (Sanborn et al. 2015). The nature of ordinary chromatin domains is less clear. It has been proposed that these domains originate due to clustering and spatial segregation of active and inactive genomic regions. Accordingly, it was proposed to call them "compartmental domains" (Rowley and Corces 2018). The mechanisms underlying the spatial segregation of chromatin compartments (or compartmental domains) are still unclear. A current model postulates that proteins enriched in different chromatin types trigger phase separation, resulting in their spatial segregation (Nuebler et al. 2018; Rada-Iglesias et al. 2018).

Functional Domains of the Eukaryotic Genome

The eukaryotic genome has long been proposed to be a mosaic of semi-independent structural-functional domains (Bodnar 1988; Goldman 1988). The original model was inspired by the results of analysis of DNaseI sensitivity of individual genes and genomic segments (Weintraub and Groudine 1976; Weintraub et al. 1981; Lawson et al. 1982; Jantzen et al. 1986). It was proposed that the entire genome is built from similarly organized structural-functional units (domains) that may be either active or repressed. The transcriptional status of the domain was thought to be controlled at the level of chromatin packaging. The model stimulated research aimed to identify regulatory elements controlling the chromatin status of genomic domains. These studies resulted in identification of domain bordering elements (insulators) (Kellum and Schedl 1991, 1992; Udvardy et al. 1986), nuclear matrix attachment regions (MARs) (Cockerill and Garrard 1986), and locus control regions (LCRs) (Forrester et al. 1987, 1990; Grosveld et al. 1987; Li et al. 1990). Although in its initial form the domain model of eukaryotic genome organization cannot account for a number of recent observations, it can be upgraded taking into account the 3D genome organization (Razin and Vassetzky 2017). Considering the necessity of juxtaposition of enhancers and promoters, one may conclude that any self-interacting chromatin domain would impose certain restrictions on enhancer action. Indeed, it has been demonstrated that, in most cases, the areas of enhancers' action are restricted to the so-called insulating neighborhoods (Sun et al. 2019), regulatory archipelagos (Montavon et al. 2011), regulatory landscapes (Spitz et al. 2003; Zuniga et al. 2004), or regulatory domains (Symmons et al. 2014). These functional genomic blocks are large (100 Kb to 1 Mb) segments of the genome within which non-related genes demonstrate similar tissue specificity of expression. Being integrated in such a domain, a reporter gene under control of a minimal promoter demonstrates a tissue-specific expression profile typical for the domain as a whole (Ruf et al. 2011; Symmons et al. 2014). Although there is still some discrepancy in the results of different authors, they all agree that insulated areas colocalize with self-interacting chromatin domains identified by Hi-C analysis, either with TADs (Montavon et al. 2011; Symmons et al. 2014) or looped domains (sub-TADs) (Sun et al. 2019).

Interestingly, TADs harboring superenhancers are preferentially insulated by boundaries possessing a particularly high insulation score (Gong et al. 2018).

Partitioning of the genome into semi-independent structural-functional domains appears important for two reasons. First, it minimizes the possibility of an off-target activity of any given enhancer. To this end, it is of note that genomic rearrangements affecting TAD boundaries frequently result in compromising gene regulation networks and development of diseases (Lupianez et al. 2015; Franke et al. 2016; Valton and Dekker 2016; Ibn-Salem et al. 2014; Vicente-Garcia et al. 2017). Second, partitioning of the genome into TADs restricts the area the enhancer should explode to find a target promoter. Correspondingly, the time necessary to establish enhancer-promoter communication is reduced (Symmons et al. 2016). Lack of rigidity in the TAD structure is of importance in this context. Alternative configurations of the chromatin fiber continuously interchange within a TAD (Tiana et al. 2016). This interchange is likely to provide additional possibilities for cell adaptation to a changing environment (Razin et al. 2013). The functional relevance of genome partitioning into TADs is likely to explain the apparent conservation of this organization in the genomes of related species (Dixon et al. 2012) as well as the fact that TADs are stable against rearrangements during evolution (Krefting et al. 2018; Lazar et al. 2018). Interestingly, paralog gene pairs are enriched for colocalization in the same TAD and frequently share common enhancer elements (Ibn-Salem et al. 2017).

Besides constituting the insulation neighborhoods for transcription regulation, the TADs also contribute to the control of replication because they correspond to units of replication timing (replication domains) (Pope et al. 2014). Interestingly, after being disrupted in mitosis (Naumova et al. 2013), TADs are re-established in G1 phase of the cell cycle at about the same time with the establishment of the replication-timing program (Dileep et al. 2015a, b). It may be that exactly at the level of chromatin packaging, the link between active transcription and early replication is established.

TAD Assembly and Insulation

Taking into consideration the fact that TADs restrict the areas of enhancer action, it is particularly important to understand how they are assembled and why they are insulated. Comparison of Hi-C maps with genome-wide distribution of various epigenetic marks demonstrated that, in mammals, TAD boundaries are enriched in CTCF-binding sites and active genes (Dixon et al. 2012). Also, cohesin was found enriched at TAD boundaries (Hansen et al. 2017). Deletion of CTCF-binding sites at TAD boundaries resulted in a full or partial loss of TAD insulation (Narendra et al. 2015, 2016; Lupianez et al. 2015; Sanborn et al. 2015). The same effect was observed upon targeted degradation of CTCF in living cells (Nora et al. 2017). CTCF has long been implicated in mediation of enhancer-blocking activity of insulators (Chung et al. 1997). In addition, it mediates formation of DNA/chromatin loops (Vietri Rudan and Hadjur 2015; Holwerda and de Laat 2012). It should be mentioned, however, that by itself, formation of a chromatin loop is not sufficient for TAD assembly. Within a loop, only the bases are permanently located in a spatial proximity. On a Hi-C heat map, a DNA loop can be recognized as a high interaction signal between bases that looks like a spot at the top of a triangle. However, to "fill" the triangle, it is necessary to ensure mutual interaction of internal parts of the loop (Fig. 2.3a). It is also not clear how deposition of CTCF at TAD boundaries can prevent spatial interactions between internal regions of different TADs. Although CTCF is a large protein (~130 kDa), the octamer of histones constituting the nucleosomal core has approximately the same summary weight, and the 1 Mb mammalian TAD is composed of ~5000 nucleosomes. It is easy to speculate about a mechanism by which deposition of CTCF can interfere with spreading if signals travel along a linear chromatin fiber. Here, a traffic jam model fits perfectly (Fig. 2.3b). However, it is difficult to see how spatial interactions between internal regions of large TADs can be prevented by CTCF (Fig. 2.3c). In fact, it is easier to consider a possibility that TAD is held together by some internal links (see below). However, preferential deposition of CTCF as well as cohesin at mammalian TAD boundaries is an established fact (Sofueva et al. 2013; Nora et al. 2012; Dixon et al. 2012; Zuin et al. 2014; Wutz et al. 2017), and there should be a reason for this deposition.

The model explaining the roles of CTCF and cohesion in TAD formation was suggested by two research teams (Fudenberg et al. 2016; Sanborn et al. 2015). According to the model, cohesin mediates DNA loop extrusion. The process of extrusion may start anywhere in the genome but cannot pass CTCF-binding sites present in a certain orientation. The last supposition was based on the observation that CTCF-binding motive has a direction and that CTCF-binding motives present at TAD boundaries (and bases of sub-TAD loops) usually have convergent orientation (Sanborn et al. 2015; Vietri Rudan et al. 2015; de Wit et al. 2015). Of note, the model considers TAD as a population phenomenon. In each individual cell, only a loop or a set of loops exist within the area that is considered as a TAD. However, all Hi-C maps that have been discussed so far were obtained when cell populations were studied. That is typical for a normal biochemical experiment. In a typical Hi-C protocol, one starts with 1–10 millions of cells. The loop extrusion model assumes that filled triangles (TADs) seen on population Hi-C maps represent superimposition of signals reflecting mainly interaction of bases of a variety of loops extruded in individual cells. This model has been supported by in silico modeling (Fudenberg et al. 2016). Also, it has been demonstrated that depletion or degrading of cohesin results in partial or full disruption of TADs (Sofueva et al. 2013; Rao et al. 2017), whereas depletion of cohesin unloading factor WAPL results in generation of longer chromatin loops (Wutz et al. 2017; Haarhuis et al. 2017) as predicted by the DNA loop extrusion model. The main challenge of the model is that the ability of cohesin to extrude DNA loops was not directly demonstrated. At the same time, it is known that cohesin possesses ATPase activity (Hirano 2005) and is able to move along

DNA both in vitro (Stigler et al. 2016; Kanke et al. 2016) and in vivo (Busslinger et al. 2017). Of note, this movement is restricted by CTCF (Davidson et al. 2016; Busslinger et al. 2017). Recently published results of Casellas's lab demonstrated that loop domains are formed by a process that requires cohesin ATPases (Vian et al. 2018). Finally, a condensin complex that is closely related to cohesin was found able to extrude DNA loops (Ganji et al. 2018). Taken together, these observations strongly support a supposition that cohesin may act as a DNA loop extrusion motor in the interphase nucleus.

It should be stressed that the DNA loop extrusion model (Fudenberg et al. 2016; Sanborn et al. 2015) considers TAD as a population phenomenon. The single-cell Hi-C studies performed so far have not provided a definitive answer to the question of whether there are TADs in individual mammalian cells due to a low resolution of Hi-C maps (Nagano et al. 2013; Flyamer et al. 2017). On the other hand, compact, and at first approximation globular, domains can be visualized in nuclei by FISH with TAD-specific probes (Bintu et al. 2018; Szabo et al. 2018). It is thus likely that there should be another mechanism that ensures compactization of entire TADs or extruded loops. It has been proposed that entropic forces primarily drive the formation of compact contact domains in a polymer confined to a limited space (Vasquez et al. 2016). This supposition made based on results of computational simulations is indirectly supported by the fact that contact domains occur in one or another form in the genomes of various organisms, including bacteria (Le et al. 2013), and special cell types, such as spermatozoa, which contain protamines in place of histones in their nuclei (Battulin et al. 2015). However, organization of nucleosomal fiber into compact domains may be also promoted by electrostatic interaction between nucleosomal particles. The ability of nucleosomal fibers to form various conglomerates is well documented. The conglomerates are stabilized by interactions between positively charged N-terminal tails of histones H3 and H4 and a negatively charged acidic patch on the surface of a nucleosomal globule (Kalashnikova et al. 2013; Pepenella et al. 2014). The same interactions facilitate the formation of 30-nm nucleosome fibers at low fiber concentrations, when between-fiber contacts are unlikely (Luger et al. 1997; Sinha and Shogren-Knaak 2010).

The main concern regarding the model of TAD assembly by condensation of nucleosomal fibers is to explain why individual TADs are separated. To this end, it should be mentioned that, in Drosophila, CTCF loops do not play a major role in 3D genome organization (Rowley et al. 2017). We and others reported that, in Drosophila cells, TAD boundaries harbor transcribed genes and are enriched in histone modifications typical for active chromatin (Ulianov et al. 2016; Sexton et al. 2012; Hou et al. 2012). Histone acetylation, which is typical of active chromatin, decreases the histone charge and prevents internucleosome interactions (Shogren-Knaak et al. 2006; Allahverdi et al. 2011). We argued that these processes may be sufficient to prevent assembly of active chromatin regions into compact domains (Ulianov et al. 2016). Thus, the distribution of active and inactive genes along a DNA molecule may determine the profile of chromosome organization in TADs. To test this idea,

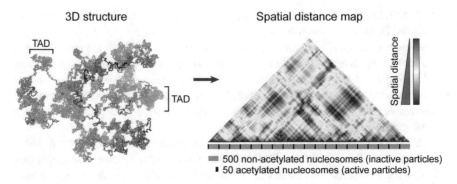

Fig. 2.4 Model heteropolymer built up from long blocks of inactive particles (non-acetylated nucleosomes interacting with each other) interspersed with short blocks of active particles (acetylated nucleosomes unable to interact with other nucleosomes) recapitulates some structural properties of chromatin. Polymer simulations demonstrate that blocks of inactive particles fold into globules manifested as TADs in spatial distance maps of the polymer. The results of a typical simulation are presented

we performed computer modelling of self-folding of a virtual polymer that consists of alternating nucleosome blocks of two types reproducing the properties of active and inactive chromatin regions (Fig. 2.4) (Ulianov et al. 2016). The particles of inactive block (500 particles in each block) were allowed to establish a limited number of relatively unstable contacts with the particles of the same type from the same or other inactive blocks. The particles of active blocks (50 particles in each block) were not allowed to establish contacts with each other or with particles from inactive blocks. The self-folding of polymer simulated using dissipative particle dynamics algorithm resulted in formation of globular structures roughly colocalizing with inactive blocks separated by unfolded active blocks (Ulianov et al. 2016). Of course, in each individual simulation, the folding of polymer was not fully regular. In some cases, conglomerates of inactive nucleosomes fused to produce superconglomerates; in other cases, nucleosomes of one inactive block formed more than one conglomerate with less compact spacers between the conglomerates (Fig. 2.4). However, averaging of the results of 12 simulations allowed generation of a Hi-C map containing contact domains (TADs) that coincided with inactive nucleosome blocks and were separated by spacers of active nucleosomes (Ulianov et al. 2016). Other simulations have demonstrated that short patches of "active chromatin" inserted into "inactive chromatin" blocks tend to be extruded on a surface of inactive block (Gavrilov et al. 2016). Insertion of larger stretches of "active chromatin" resulted rather in splitting of inactive blocks. This observation was in agreement with experimental observations that activation of transcription of tissue-specific genes located within TADs correlates with decompacting of the corresponding region, which, in some cases, resulted in TAD splitting (Ulianov et al. 2016).

It should be mentioned that DNA loop extrusion and nucleosome condensation are not mutually exclusive. Thus, nucleosome condensation may contribute to the compaction of extruded chromatin loops in mammalian cells. There is yet another group of models postulating that TAD formation is mediated by architectural proteins that form intra-TAD links, thus pulling together remote segments of a chromatin fiber. To explain the existence of isolated TADs, the models assume a multiplicity of architectural protein groups, each ensuring the formation of a particular TAD (Barbieri et al. 2012, 2013; Pombo and Nicodemi 2014). The models are supported by computer simulations but seem implausible biologically because there are 100 times fewer architectural protein types than TADs even in Drosophila, which is known to have several architectural proteins in addition to CTCF (Zolotarev et al. 2016).

3D Organization of the Genome in the Context of Nuclear Compartmentalization

The current model of the global genome organization within the eukaryotic cell nucleus was formulated long before the development of Hi-C and other C-methods. Initially, this model was based exclusively on the results of microscopic studies. Territorial organization of interphase chromosomes and the existence of an interchromatin domain (ICD) that spans chromosomal territories are the main points of the model (Cremer and Cremer 2001, 2010, 2018; Cremer et al. 2017, 2018). The interchromatin domain is the place where various membraneless nuclear bodies such as nucleoli, splicing speckles, Cajal bodies, paraspeckles, histone locus bodies, and PML bodies are assembled (for a review, see Mao et al. (2011); Ulianov et al. (2015); Stanek and Fox (2017)). The initial version of the model placed ICD between chromosomal territories (Cremer et al. 1993; Zirbel et al. 1993). With the increase of resolution of microscopic methods, it became evident that the ICD also penetrates chromosomal territories (Cremer and Cremer 2010, 2018). Chromosome territories themselves are composed of chromatin domains and chromatin domain clusters that likely correspond to TADs and contact domains of higher order. Interestingly, internal parts of these domains appear to contain mostly inactive chromatin, whereas active genes are preferentially located at the perichromatin layer (Cremer and Cremer 2018; Cremer et al. 2018). Although individual chromosomes constitute rather separated entities within the cell nucleus, interchromosomal contacts could still be found at various reaction centers such as transcription factories, PML bodies, and splicing speckles. Such contacts were first observed using FISH to visualize various genes in combination with immunostaining to observe functional nuclear compartments (Wang et al. 2004; Sun et al. 2003; Shopland et al. 2003; Szczerbal and Bridger 2010; Moen et al. 2004) and then reanalyzed using genome-wide C-methods (Wang et al. 2016; Schoenfelder et al. al. 2010; Quinodoz et al. 2018).

It should be mentioned that biochemical protocols based on a proximity ligation (C-methods) allow for identification of only particularly close spatial contacts. Recruitment of several genomic regions to the same compartment is difficult, if not impossible, to detect using C-methods. Development of alternative experimental protocols based on barcoding of DNA fragments present within the same, even quite large, fixed chromatin complex (Quinodoz et al. 2018) solved the problem. Using such an experimental procedure termed "SPRITE" (split-pool recognition of interactions by tag extension), Quinodoz et al. have identified two hubs of interchromosomal interactions that are arranged around the nucleolus (repressed hub) and nuclear speckles (active hub) (Quinodoz et al. 2018). Another genome-wide protocol that enables measuring distances between various genes and nuclear compartments is TSA-Seq (Chen et al. 2018). The procedure utilizes the tyramide amplification cascade (Wang et al. 1999) to biotinylate DNA in the vicinity of sites to which horseradish peroxidase (HRP) catalyzes the formation of tyramide-biotin free radicals recruited using an appropriate cascade of antibodies. Biotinylated DNA is then pulled down on streptavidin and sequenced. Using TSA-Seq, Belmont and coauthors confirmed clustering of active genes close to nuclear speckles. In agreement with a number of previous reports (Shevelyov and Nurminsky 2012; van Steensel and Belmont 2017), the repressed genes were found more in proximity to the nuclear lamina (Chen et al. 2018).

Taking together, the above observations argue that 3D organization of the genome and functional compartmentalization of the cell nucleus are mutually dependent. 3D organization is not simply a sum of enhancer-promoter and CTCF loops. It relies on a number of factors present in non-disturbed nuclei. Various fractionation procedures compromise this complex organization and drastically affect the results of analysis based on capturing pairwise interactions of remote DNA fragments (Gavrilov et al. 2013). Juxtaposition of remote genomic elements is not only ensured by interaction of proteins bound to these elements but rather represents a result of specific folding of a large genomic segment supported by numerous interactions outside the juxtaposed regions (Razin et al. 2013). These interactions include repositioning of various genomic segments to the vicinity of functional nuclear compartments. On the other hand, the folded genome as a whole provides a structural basis for nuclear compartmentalization (Misteli 2007; Schneider and Grosschedl 2007; Lanctot et al. 2007; Razin et al. 2013). The ICD where all these compartments are assembled is formed by exclusion from the areas occupied by chromatin. Segregation of interphase chromosomes resulting in the existence of chromosomal territories appears to be ensured by basic physical properties of charged polymers (Rosa and Everaers 2008; Mateos-Langerak et al. 2009; Bohn and Heermann 2010; Tark-Dame et al. 2011). It is less clear what supports the existence of channeled compartment within chromosomal territories. The simplest supposition is that repulsion between surfaces of TADs is of primary importance. The key point to be taken into account is that the surface of TADs should be more charged than the internal regions. Recent results of the Cremer team demonstrate that active chromatin is located at the surface of 1 Mb chromatin domains (TADs) (Cremer and Cremer

2018; Cremer et al. 2018) and thus lines the ICD channels. This finding is corroborated by the results of in silico modeling of TAD assembly (Gavrilov et al. 2016). High levels of histone acetylation typical for active chromatin (Shogren-Knaak et al. 2006; Allahverdi et al. 2011) should make the perichromatin layer more negatively charged compared to the internal part of chromatin domains/TADs. Thus, the perichromatin layer should stabilize and insulate inactive chromatin domains/TADs via generating electrostatic repulsion between them. This layer may prevent intermingling of TADs and ensure existence of intrachromosomal channels. The basic landscape for nuclear compartmentalization is thus directed only by physical laws (Rosa and Everaers 2008; Cook and Marenduzzo 2009; Dorier and Stasiak 2009; Kim and Szleifer 2014). Once established after mitosis, the territorial organization of interphase chromosomes becomes stabilized by interaction of certain chromosomal regions with the nuclear lamina (Guelen et al. 2008; Pickersgill et al. 2006) and nucleolus (Nemeth et al. 2010; van Koningsbruggen et al. 2010). Nucleoli are assembled at particular genomic loci harboring arrays of rRNA genes. The same is true for histone locus bodies. Transcription factories are likely to assemble stochastically by aggregation of closely located transcription complexes (Razin et al. 2011). Still, spatial positioning of the involved transcribed genes will predetermine their location. Typically for biological systems, this organization is highly dynamic. This dynamism applies to the both folding of interphase chromosomes and assembly of nuclear compartments. Live imaging studies have demonstrated that both chromosome territories and individual domains within chromosomal territories undergo constant movement (Marshall et al. 1997a, b, Marshall 2002; Levi et al. 2005; Pliss et al. 2013). The typical configuration of an interphase chromosome or shorter genomic segments represents an equilibrium of a number of possible configurations (Nagano et al. 2013; Stevens et al. 2017). The nature of functional nuclear compartments has been a matter of long-term discussions. The current model suggests that these compartments are liquid droplets formed by phase separation. They can fuse or separate into smaller droplets depending on external conditions. Although each type of compartments is rich in a particular set of proteins, the sets of proteins present in different compartments may overlap, and proteins present within compartments rapidly exchange with those proteins present in nucleoplasm. Furthermore, while speckles were reported to be positionally stable within hours (Misteli et al. 1997; Kruhlak et al. 2000), Cajal bodies and PML bodies appear to diffuse within the ICD as freely as an artificially created inert object of the same dimensions (Gorisch et al. 2004). An apparent order within the cell nucleus is thus likely to emerge out of a disorder due to a shaky equilibrium of different forces including a depletion attraction force (Cho and Kim 2012; Marenduzzo et al. 2006; Hancock 2004b; Rippe 2007). Apparently, the interplay between various functional processes that occur in the nucleus in any given moment directs both the chromosome folding and spatial compartmentalization of the nucleus (Rippe 2007; Kim and Szleifer 2014; Hancock 2004a; Razin et al. 2013; Golov et al. 2015; Sengupta 2018; Shah et al. 2018). Consequently, the cell nucleus should be considered as an integrated system, the properties of which emerge due to the interaction of numerous components and cannot be fully explained or predicted based on the properties of individual components. Further progress in understanding mechanisms of eukaryotic genome functioning will depend on reconsideration of all pull of existing data in terms of systems biology.

Acknowledgments This work was supported by the Russian Science Foundation (grant 19-14-00016) and by the Interdisciplinary Scientific and Educational School of Moscow University «Molecular Technologies of the Living Systems and Synthetic Biology».

References

Allahverdi A, Yang R, Korolev N, Fan Y, Davey CA, Liu CF, Nordenskiold L (2011) The effects of histone H4 tail acetylations on cation-induced chromatin folding and self-association. Nucleic Acids Res 39(5):1680–1691. https://doi.org/10.1093/nar/gkq900

Arnold CD, Gerlach D, Stelzer C, Boryn LM, Rath M, Stark A (2013) Genome-wide quantitative enhancer activity maps identified by STARR-seq. Science 339(6123):1074–1077. https://doi.org/10.1126/science.1232542

Barbieri M, Chotalia M, Fraser J, Lavitas LM, Dostie J, Pombo A, Nicodemi M (2012) Complexity of chromatin folding is captured by the strings and binders switch model. Proc Natl Acad Sci U S A 109(40):16173–16178. https://doi.org/10.1073/pnas.1204799109

Barbieri M, Fraser J, Lavitas LM, Chotalia M, Dostie J, Pombo A, Nicodemi M (2013) A polymer model explains the complexity of large-scale chromatin folding. Nucleus 4(4):267–273

Battulin N, Fishman VS, Mazur AM, Pomaznoy M, Khabarova AA, Afonnikov DA, Prokhortchouk EB, Serov OL (2015) Comparison of the three-dimensional organization of sperm and fibroblast genomes using the Hi-C approach. Genome Biol 16:77. https://doi.org/10.1186/s13059-015-0642-0

Bintu B, Mateo LJ, Su JH, Sinnott-Armstrong NA, Parker M, Kinrot S, Yamaya K, Boettiger AN, Zhuang X (2018) Super-resolution chromatin tracing reveals domains and cooperative interactions in single cells. Science 362(6413). https://doi.org/10.1126/science.aau1783

Bodnar JW (1988) A domain model for eukaryotic DNA organization: a molecular basis for cell differentiation and chromosome evolution. J Theor Biol 132(4):479–507

Bohn M, Heermann DW (2010) Diffusion-driven looping provides a consistent framework for chromatin organization. PLoS One 5(8):e12218. https://doi.org/10.1371/journal.pone.0012218

Bondarenko VA, Liu YV, Jiang YI, Studitsky VM (2003) Communication over a large distance: enhancers and insulators. Biochem Cell Biol 81(3):241–251. https://doi.org/10.1139/o03-051

Bonev B, Cavalli G (2016) Organization and function of the 3D genome. Nat Rev Genet 17(11):661–678. https://doi.org/10.1038/nrg.2016.112

Busslinger GA, Stocsits RR, van der Lelij P, Axelsson E, Tedeschi A, Galjart N, Peters JM (2017) Cohesin is positioned in mammalian genomes by transcription, CTCF and Wapl. Nature 544(7651):503–507. https://doi.org/10.1038/nature22063

Chen Y, Zhang Y, Wang Y, Zhang L, Brinkman EK, Adam SA, Goldman R, van Steensel B, Ma J, Belmont AS (2018) Mapping 3D genome organization relative to nuclear compartments using TSA-Seq as a cytological ruler. J Cell Biol 217(11):4025–4048. https://doi.org/10.1083/jcb.201807108

Cho EJ, Kim JS (2012) Crowding effects on the formation and maintenance of nuclear bodies: insights from molecular-dynamics simulations of simple spherical model particles. Biophys J 103(3):424–433. https://doi.org/10.1016/j.bpj.2012.07.007

Chung JH, Bell AC, Felsenfeld G (1997) Characterization of the chicken beta-globin insulator. Proc Natl Acad Sci U S A 94(2):575–580

Cockerill PN, Garrard WT (1986) Chromosomal loop anchorage of the kappa immunoglobulin gene occurs next to the enhancer in a region containing topoisomerase II sites. Cell 44:273–282

Consortium EP, Bernstein BE, Birney E, Dunham I, Green ED, Gunter C, Snyder M (2012) An integrated encyclopedia of DNA elements in the human genome. Nature 489(7414):57–74. https://doi.org/10.1038/nature11247

Cook PR, Marenduzzo D (2009) Entropic organization of interphase chromosomes. J Cell Biol 186(6):825–834. https://doi.org/10.1083/jcb.200903083

Cremer T, Cremer C (2001) Chromosome territories, nuclear architecture and gene regulation in mammalian cells. Nat Rev Genet 2(4):292–301

Cremer T, Cremer M (2010) Chromosome territories. Cold Spring Harb Perspect Biol 2(3):a003889. https://doi.org/10.1101/cshperspect.a003889

Cremer M, Cremer T (2018) Nuclear compartmentalization, dynamics, and function of regulatory DNA sequences. Genes Chromosomes Cancer. https://doi.org/10.1002/gcc.22714

Cremer T, Kurz A, Zirbel R, Dietzel S, Rinke B, Schrock E, Speicher MR, Mathieu U, Jauch A, Emmerich P, Scherthan H, Ried T, Cremer C, Lichter P (1993) Role of chromosome territories in the functional compartmentalization of the cell nucleus. Cold Spring Harb Symp Quant Biol 58:777–792

Cremer M, Schmid VJ, Kraus F, Markaki Y, Hellmann I, Maiser A, Leonhardt H, John S, Stamatoyannopoulos J, Cremer T (2017) Initial high-resolution microscopic mapping of active and inactive regulatory sequences proves non-random 3D arrangements in chromatin domain clusters. Epigenetics Chromatin 10(1):39. https://doi.org/10.1186/s13072-017-0146-0

Cremer T, Cremer M, Cremer C (2018) The 4D nucleome: genome compartmentalization in an evolutionary context. Biochemistry (Mosc) 83(4):313–325. https://doi.org/10.1134/S000629791804003X

Davidson IF, Goetz D, Zaczek MP, Molodtsov MI, Huis In't Veld PJ, Weissmann F, Litos G, Cisneros DA, Ocampo-Hafalla M, Ladurner R, Uhlmann F, Vaziri A, Peters JM (2016) Rapid movement and transcriptional re-localization of human cohesin on DNA. EMBO J 35(24):2671–2685. https://doi.org/10.15252/embj.201695402

de Laat W, Grosveld F (2003) Spatial organization of gene expression: the active chromatin hub. Chromosome Res 11:447–459

de Wit E, de Laat W (2012) A decade of 3C technologies: insights into nuclear organization. Genes Dev 26(1):11–24

de Wit E, Vos ES, Holwerda SJ, Valdes-Quezada C, Verstegen MJ, Teunissen H, Splinter E, Wijchers PJ, Krijger PH, de Laat W (2015) CTCF binding polarity determines chromatin looping. Mol Cell 60(4):676–684. https://doi.org/10.1016/j.molcel.2015.09.023

Dekker J, Mirny L (2016) The 3D genome as moderator of chromosomal communication. Cell 164(6):1110–1121. https://doi.org/10.1016/j.cell.2016.02.007

Dekker J, Rippe K, Dekker M, Kleckner N (2002) Capturing chromosome conformation. Science 295(5558):1306–1311

Denker A, de Laat W (2016) The second decade of 3C technologies: detailed insights into nuclear organization. Genes Dev 30(12):1357–1382. https://doi.org/10.1101/gad.281964.116

Dileep V, Ay F, Sima J, Vera DL, Noble WS, Gilbert DM (2015a) Topologically associating domains and their long-range contacts are established during early G1 coincident with the establishment of the replication-timing program. Genome Res 25(8):1104–1113. https://doi.org/10.1101/gr.183699.114

Dileep V, Rivera-Mulia JC, Sima J, Gilbert DM (2015b) Large-scale chromatin structure-function relationships during the cell cycle and development: insights from replication timing. Cold Spring Harb Symp Quant Biol. https://doi.org/10.1101/sqb.2015.80.027284

Dixon JR, Selvaraj S, Yue F, Kim A, Li Y, Shen Y, Hu M, Liu JS, Ren B (2012) Topological domains in mammalian genomes identified by analysis of chromatin interactions. Nature 485(7398):376–380. https://doi.org/10.1038/nature11082

Dixon JR, Gorkin DU, Ren B (2016) Chromatin domains: the unit of chromosome organization. Mol Cell 62(5):668–680. https://doi.org/10.1016/j.molcel.2016.05.018

Dorier J, Stasiak A (2009) Topological origins of chromosomal territories. Nucleic Acids Res 37(19):6316–6322. https://doi.org/10.1093/nar/gkp702

Eagen KP (2018) Principles of chromosome architecture revealed by Hi-C. Trends Biochem Sci 43(6):469–478. https://doi.org/10.1016/j.tibs.2018.03.006

Eser U, Chandler-Brown D, Ay F, Straight AF, Duan Z, Noble WS, Skotheim JM (2017) Form and function of topologically associating genomic domains in budding yeast. Proc Natl Acad Sci U S A 114(15):E3061–E3070. https://doi.org/10.1073/pnas.1612256114

Flyamer IM, Gassler J, Imakaev M, Brandao HB, Ulianov SV, Abdennur N, Razin SV, Mirny LA, Tachibana-Konwalski K (2017) Single-nucleus Hi-C reveals unique chromatin reorganization at oocyte-to-zygote transition. Nature 544(7648):110–114. https://doi.org/10.1038/nature21711

Forrester WC, Takegawa S, Papayannopoulou T, Stamatoyannopoulos G, Groudine M (1987) Evidence for a locus activating region: the formation of developmentally stable hypersensitive sites in globin expressing hybrids. Nucleic Acids Res 15:10159–10175

Forrester WC, Epner E, Driscoll MC, Enver T, Brice M, Papayannopoulou T, Groudine M (1990) A deletion of the human b-globin locus activation region causes a major alteration in chromatin structure and replication across the entire b-globin locus. Gene Dev 4:1637–1649

Franke M, Ibrahim DM, Andrey G, Schwarzer W, Heinrich V, Schopflin R, Kraft K, Kempfer R, Jerkovic I, Chan WL, Spielmann M, Timmermann B, Wittler L, Kurth I, Cambiaso P, Zuffardi O, Houge G, Lambie L, Brancati F, Pombo A, Vingron M, Spitz F, Mundlos S (2016) Formation of new chromatin domains determines pathogenicity of genomic duplications. Nature 538(7624):265–269. https://doi.org/10.1038/nature19800

Fraser J, Ferrai C, Chiariello AM, Schueler M, Rito T, Laudanno G, Barbieri M, Moore BL, Kraemer DC, Aitken S, Xie SQ, Morris KJ, Itoh M, Kawaji H, Jaeger I, Hayashizaki Y, Carninci P, Forrest AR, Consortium F, Semple CA, Dostie J, Pombo A, Nicodemi M (2015) Hierarchical folding and reorganization of chromosomes are linked to transcriptional changes in cellular differentiation. Mol Syst Biol 11(12):852. https://doi.org/10.15252/msb.20156492

Fudenberg G, Imakaev M, Lu C, Goloborodko A, Abdennur N, Mirny LA (2016) Formation of chromosomal domains by loop extrusion. Cell Rep 15(9):2038–2049. https://doi.org/10.1016/j.celrep.2016.04.085

Fullwood MJ, Liu MH, Pan YF, Liu J, Xu H, Mohamed YB, Orlov YL, Velkov S, Ho A, Mei PH, Chew EG, Huang PY, Welboren WJ, Han Y, Ooi HS, Ariyaratne PN, Vega VB, Luo Y, Tan PY, Choy PY, Wansa KD, Zhao B, Lim KS, Leow SC, Yow JS, Joseph R, Li H, Desai KV, Thomsen JS, Lee YK, Karuturi RK, Herve T, Bourque G, Stunnenberg HG, Ruan X, Cacheux-Rataboul V, Sung WK, Liu ET, Wei CL, Cheung E, Ruan Y (2009) An oestrogen-receptor-alpha-bound human chromatin interactome. Nature 462(7269):58–64

Fussner E, Strauss M, Djuric U, Li R, Ahmed K, Hart M, Ellis J, Bazett-Jones DP (2012) Open and closed domains in the mouse genome are configured as 10-nm chromatin fibres. EMBO Rep 13(11):992–996. https://doi.org/10.1038/embor.2012.139

Ganji M, Shaltiel IA, Bisht S, Kim E, Kalichava A, Haering CH, Dekker C (2018) Real-time imaging of DNA loop extrusion by condensin. Science 360(6384):102–105. https://doi.org/10.1126/science.aar7831

Gavrilov AA, Razin SV (2008) Spatial configuration of the chicken {alpha}-globin gene domain: immature and active chromatin hubs. Nucleic Acids Res 36:4629–4640

Gavrilov AA, Gushchanskaya ES, Strelkova O, Zhironkina O, Kireev II, Iarovaia OV, Razin SV (2013) Disclosure of a structural milieu for the proximity ligation reveals the elusive nature of an active chromatin hub. Nucleic Acids Res 41:3563–3575. https://doi.org/10.1093/nar/gkt067

Gavrilov AA, Shevelyov YY, Ulianov SV, Khrameeva EE, Kos P, Chertovich A, Razin SV (2016) Unraveling the mechanisms of chromatin fibril packaging. Nucleus 7(3):319–324. https://doi.org/10.1080/19491034.2016.1190896

Getzenberg RH, Pienta KJ, Ward WS, Coffey DS (1991) Nuclear structure and the three-dimensional organization of DNA. J Cell Biochem 47(4):289–299. https://doi.org/10.1002/jcb.240470402

Ghavi-Helm Y, Klein FA, Pakozdi T, Ciglar L, Noordermeer D, Huber W, Furlong EE (2014) Enhancer loops appear stable during development and are associated with paused polymerase. Nature 512(7512):96–100. https://doi.org/10.1038/nature13417

Gibcus JH, Dekker J (2013) The hierarchy of the 3D genome. Mol Cell 49(5):773–782. https://doi.org/10.1016/j.molcel.2013.02.011

Goldman MA (1988) The chromatin domain as a unit of gene regulation. BioEssays 9:50–55

Golov AK, Gavrilov AA, Razin SV (2015) The role of crowding forces in juxtaposing beta-globin gene domain remote regulatory elements in mouse erythroid cells. PLoS One 10(10):e0139855. https://doi.org/10.1371/journal.pone.0139855

Gong Y, Lazaris C, Sakellaropoulos T, Lozano A, Kambadur P, Ntziachristos P, Aifantis I, Tsirigos A (2018) Stratification of TAD boundaries reveals preferential insulation of super-enhancers by strong boundaries. Nat Commun 9(1):542. https://doi.org/10.1038/s41467-018-03017-1

Gorisch SM, Wachsmuth M, Ittrich C, Bacher CP, Rippe K, Lichter P (2004) Nuclear body movement is determined by chromatin accessibility and dynamics. Proc Natl Acad Sci U S A 101(36):13221–13226. https://doi.org/10.1073/pnas.0402958101

Grosveld F, van Assandelt GB, Greaves DR, Kollias B (1987) Position-independent, high-level expression of the human b-globin gene in transgenic mice. Cell 51:975–985

Guelen L, Pagie L, Brasset E, Meuleman W, Faza MB, Talhout W, Eussen BH, de Klein A, Wessels L, de Laat W, van Steensel B (2008) Domain organization of human chromosomes revealed by mapping of nuclear lamina interactions. Nature 453(7197):948–951. https://doi.org/10.1038/nature06947

Haarhuis JHI, van der Weide RH, Blomen VA, Yanez-Cuna JO, Amendola M, van Ruiten MS, Krijger PHL, Teunissen H, Medema RH, van Steensel B, Brummelkamp TR, de Wit E, Rowland BD (2017) The cohesin release factor WAPL restricts chromatin loop extension. Cell 169 (4):693-707. e614. doi:https://doi.org/10.1016/j.cell.2017.04.013

Hancock R (2004a) Internal organisation of the nucleus: assembly of compartments by macromolecular crowding and the nuclear matrix model. Biol Cell 96(8):595–601

Hancock R (2004b) A role for macromolecular crowding effects in the assembly and function of compartments in the nucleus. J Struct Biol 146(3):281–290. https://doi.org/10.1016/j.jsb.2003.12.008

Hancock R (2008) Self-association of polynucleosome chains by macromolecular crowding. Eur Biophys J: EBJ 37(6):1059–1064. https://doi.org/10.1007/s00249-008-0276-1

Hansen AS, Pustova I, Cattoglio C, Tjian R, Darzacq X (2017) CTCF and cohesin regulate chromatin loop stability with distinct dynamics. eLife 6. https://doi.org/10.7554/eLife.25776

Hirano T (2005) SMC proteins and chromosome mechanics: from bacteria to humans. Philos Trans R Soc Lond B Biol Sci 360(1455):507–514. https://doi.org/10.1098/rstb.2004.1606

Holwerda S, de Laat W (2012) Chromatin loops, gene positioning, and gene expression. Front Genet 3:217. https://doi.org/10.3389/fgene.2012.00217

Hou C, Li L, Qin ZS, Corces VG (2012) Gene density, transcription, and insulators contribute to the partition of the Drosophila genome into physical domains. Mol Cell 48(3):471–484. https://doi.org/10.1016/j.molcel.2012.08.031

Hsieh TH, Weiner A, Lajoie B, Dekker J, Friedman N, Rando OJ (2015) Mapping nucleosome resolution chromosome folding in yeast by micro-C. Cell 162(1):108–119. https://doi.org/10.1016/j.cell.2015.05.048

Ibn-Salem J, Kohler S, Love MI, Chung HR, Huang N, Hurles ME, Haendel M, Washington NL, Smedley D, Mungall CJ, Lewis SE, Ott CE, Bauer S, Schofield PN, Mundlos S, Spielmann M, Robinson PN (2014) Deletions of chromosomal regulatory boundaries are associated with congenital disease. Genome Biol 15(9):423. https://doi.org/10.1186/s13059-014-0423-1

Ibn-Salem J, Muro EM, Andrade-Navarro MA (2017) Co-regulation of paralog genes in the three-dimensional chromatin architecture. Nucleic Acids Res 45(1):81–91. https://doi.org/10.1093/nar/gkw813

Jantzen K, Friton HP, Igo-Kimenes T (1986) The DNase I sensitive domain of the chicken lyzozyme gene spans 24 kb. Nucleic Acids Res 14:6085–6099

Jin F, Li Y, Dixon JR, Selvaraj S, Ye Z, Lee AY, Yen CA, Schmitt AD, Espinoza CA, Ren B (2013) A high-resolution map of the three-dimensional chromatin interactome in human cells. Nature 503(7475):290–294. https://doi.org/10.1038/nature12644

Kalashnikova AA, Porter-Goff ME, Muthurajan UM, Luger K, Hansen JC (2013) The role of the nucleosome acidic patch in modulating higher order chromatin structure. J R Soc, Interface/ R Soc 10(82):20121022. https://doi.org/10.1098/rsif.2012.1022

Kanke M, Tahara E, Huis In't Veld PJ, Nishiyama T (2016) Cohesin acetylation and Wapl-Pds5 oppositely regulate translocation of cohesin along DNA. EMBO J 35(24):2686–2698. https://doi.org/10.15252/embj.201695756

Kellum R, Schedl P (1991) A position-effect assay for boundaries of higher-order chromosomal domains. Cell 64:941–950

Kellum R, Schedl P (1992) A group of scs elements function as boundaries in enhancer-blocking assay. Mol Cell Biol 12:2424–2431

Kim JS, Szleifer I (2014) Crowding-induced formation and structural alteration of nuclear compartments: insights from computer simulations. Int Rev Cell Mol Biol 307:73–108. https://doi.org/10.1016/B978-0-12-800046-5.00004-7

Krefting J, Andrade-Navarro MA, Ibn-Salem J (2018) Evolutionary stability of topologically associating domains is associated with conserved gene regulation. BMC Biol 16(1):87. https://doi.org/10.1186/s12915-018-0556-x

Krijger PH, de Laat W (2016) Regulation of disease-associated gene expression in the 3D genome. Nat Rev Mol Cell Biol 17(12):771–782. https://doi.org/10.1038/nrm.2016.138

Kruhlak MJ, Lever MA, Fischle W, Verdin E, Bazett-Jones DP, Hendzel MJ (2000) Reduced mobility of the alternate splicing factor (ASF) through the nucleoplasm and steady state speckle compartments. J Cell Biol 150(1):41–51

Krumm A, Duan Z (2018) Understanding the 3D genome: emerging impacts on human disease. Semin Cell Dev Biol. https://doi.org/10.1016/j.semcdb.2018.07.004

Lanctot C, Cheutin T, Cremer M, Cavalli G, Cremer T (2007) Dynamic genome architecture in the nuclear space: regulation of gene expression in three dimensions. Nat Rev Genet 8(2):104–115

Lawson GM, Knoll BJ, March CJ, Woo SLC, Tsai M-J, O'Malley BW (1982) Definition of 5' and 3' structural boundaries of the chromatin domain containing the ovalbumin multigene family. J Biol Chem 257:1501–1507

Lazar NH, Nevonen KA, O'Connell B, McCann C, O'Neill RJ, Green RE, Meyer TJ, Okhovat M, Carbone L (2018) Epigenetic maintenance of topological domains in the highly rearranged gibbon genome. Genome Res 28(7):983–997. https://doi.org/10.1101/gr.233874.117

Le TB, Imakaev MV, Mirny LA, Laub MT (2013) High-resolution mapping of the spatial organization of a bacterial chromosome. Science 342(6159):731–734. https://doi.org/10.1126/science.1242059

Levi V, Ruan Q, Plutz M, Belmont AS, Gratton E (2005) Chromatin dynamics in interphase cells revealed by tracking in a two-photon excitation microscope. Biophys J 89(6):4275–4285. https://doi.org/10.1529/biophysj.105.066670

Li Q, Zhou B, Powers P, Enver T, Stamatoyannopoulos G (1990) b-globin locus activations regions: conservation of organization, structure and function. Proc Natl Acad Sci USA 87:8207–8211

Lieberman-Aiden E, van Berkum NL, Williams L, Imakaev M, Ragoczy T, Telling A, Amit I, Lajoie BR, Sabo PJ, Dorschner MO, Sandstrom R, Bernstein B, Bender MA, Groudine M, Gnirke A, Stamatoyannopoulos J, Mirny LA, Lander ES, Dekker J (2009) Comprehensive mapping of long-range interactions reveals folding principles of the human genome. Science 326(5950):289–293

Luger K, Mader AW, Richmond RK, Sargent DF, Richmond TJ (1997) Crystal structure of the nucleosome core particle at 2.8 A resolution. Nature 389:251–260

Lupianez DG, Kraft K, Heinrich V, Krawitz P, Brancati F, Klopocki E, Horn D, Kayserili H, Opitz JM, Laxova R, Santos-Simarro F, Gilbert-Dussardier B, Wittler L, Borschiwer M, Haas SA, Osterwalder M, Franke M, Timmermann B, Hecht J, Spielmann M, Visel A, Mundlos S (2015)

Disruptions of topological chromatin domains cause pathogenic rewiring of gene-enhancer interactions. Cell 161(5):1012–1025. https://doi.org/10.1016/j.cell.2015.04.004

Lupianez DG, Spielmann M, Mundlos S (2016) Breaking TADs: how alterations of chromatin domains result in disease. Trends Genet 32(4):225–237. https://doi.org/10.1016/j.tig.2016.01.003

Luzhin AV, Flyamer IM, Khrameeva EE, Ulianov SV, Razin SV, Gavrilov AA (2019) Quantitative differences in TAD border strength underly the TAD hierarchy in Drosophila chromosomes. J Cell Biochem 120(3):4494–4503. https://doi.org/10.1002/jcb.27737

Maeshima K, Imai R, Hikima T, Joti Y (2014a) Chromatin structure revealed by X-ray scattering analysis and computational modeling. Methods 70(2-3):154–161. https://doi.org/10.1016/j.ymeth.2014.08.008

Maeshima K, Imai R, Tamura S, Nozaki T (2014b) Chromatin as dynamic 10-nm fibers. Chromosoma 123(3):225–237. https://doi.org/10.1007/s00412-014-0460-2

Maeshima K, Rogge R, Tamura S, Joti Y, Hikima T, Szerlong H, Krause C, Herman J, Seidel E, DeLuca J, Ishikawa T, Hansen JC (2016) Nucleosomal arrays self-assemble into supramolecular globular structures lacking 30-nm fibers. EMBO J. https://doi.org/10.15252/embj.201592660

Mao YS, Zhang B, Spector DL (2011) Biogenesis and function of nuclear bodies. Trends Genet 27(8):295–306

Marenduzzo D, Finan K, Cook PR (2006) The depletion attraction: an underappreciated force driving cellular organization. J Cell Biol 175(5):681–686. https://doi.org/10.1083/jcb.200609066

Markaki Y, Gunkel M, Schermelleh L, Beichmanis S, Neumann J, Heidemann M, Leonhardt H, Eick D, Cremer C, Cremer T (2010) Functional nuclear organization of transcription and DNA replication: a topographical marriage between chromatin domains and the interchromatin compartment. Cold Spring Harb Symp Quant Biol 75:475–492. https://doi.org/10.1101/sqb.2010.75.042

Markaki Y, Smeets D, Fiedler S, Schmid VJ, Schermelleh L, Cremer T, Cremer M (2012) The potential of 3D-FISH and super-resolution structured illumination microscopy for studies of 3D nuclear architecture: 3D structured illumination microscopy of defined chromosomal structures visualized by 3D (immuno)-FISH opens new perspectives for studies of nuclear architecture. BioEssays 34(5):412–426. https://doi.org/10.1002/bies.201100176

Marshall WF (2002) Order and disorder in the nucleus. Curr Biol 12(5):R185–R192

Marshall WF, Fung JC, Sedat JW (1997a) Deconstructing the nucleus: global architecture from local interactions. Curr Opin Genet Dev 7(2):259–263

Marshall WF, Straight A, Marko JF, Swedlow J, Dernburg A, Belmont A, Murray AW, Agard DA, Sedat JW (1997b) Interphase chromosomes undergo constrained diffusional motion in living cells. Curr Biol 7(12):930–939

Mateos-Langerak J, Bohn M, de Leeuw W, Giromus O, Manders EM, Verschure PJ, Indemans MH, Gierman HJ, Heermann DW, van Driel R, Goetze S (2009) Spatially confined folding of chromatin in the interphase nucleus. Proc Natl Acad Sci U S A 106(10):3812–3817. https://doi.org/10.1073/pnas.0809501106

Misteli T (2007) Beyond the sequence: cellular organization of genome function. Cell 128(4):787–800

Misteli T, Caceres JF, Spector DL (1997) The dynamics of a pre-mRNA splicing factor in living cells. Nature 387(6632):523–527. https://doi.org/10.1038/387523a0

Moen PT Jr, Johnson CV, Byron M, Shopland LS, de la Serna IL, Imbalzano AN, Lawrence JB (2004) Repositioning of muscle-specific genes relative to the periphery of SC-35 domains during skeletal myogenesis. Mol Biol Cell 15(1):197–206. https://doi.org/10.1091/mbc.E03-06-0388

Montavon T, Soshnikova N, Mascrez B, Joye E, Thevenet L, Splinter E, de Laat W, Spitz F, Duboule D (2011) A regulatory archipelago controls Hox genes transcription in digits. Cell 147(5):1132–1145. https://doi.org/10.1016/j.cell.2011.10.023

Nagano T, Lubling Y, Stevens TJ, Schoenfelder S, Yaffe E, Dean W, Laue ED, Tanay A, Fraser P (2013) Single-cell Hi-C reveals cell-to-cell variability in chromosome structure. Nature 502(7469):59–64. https://doi.org/10.1038/nature12593

Narendra V, Rocha PP, An D, Raviram R, Skok JA, Mazzoni EO, Reinberg D (2015) CTCF establishes discrete functional chromatin domains at the Hox clusters during differentiation. Science 347(6225):1017–1021. https://doi.org/10.1126/science.1262088

Narendra V, Bulajic M, Dekker J, Mazzoni EO, Reinberg D (2016) CTCF-mediated topological boundaries during development foster appropriate gene regulation. Genes Dev 30(24):2657–2662. https://doi.org/10.1101/gad.288324.116

Naumova N, Imakaev M, Fudenberg G, Zhan Y, Lajoie BR, Mirny LA, Dekker J (2013) Organization of the mitotic chromosome. Science 342(6161):948–953. https://doi.org/10.1126/science.1236083

Nemeth A, Conesa A, Santoyo-Lopez J, Medina I, Montaner D, Peterfia B, Solovei I, Cremer T, Dopazo J, Langst G (2010) Initial genomics of the human nucleolus. PLoS Genet 6(3):e1000889. https://doi.org/10.1371/journal.pgen.1000889

Nikolaou C (2017) Invisible cities: segregated domains in the yeast genome with distinct structural and functional attributes. Curr Genet. https://doi.org/10.1007/s00294-017-0731-6

Nora EP, Lajoie BR, Schulz EG, Giorgetti L, Okamoto I, Servant N, Piolot T, van Berkum NL, Meisig J, Sedat J, Gribnau J, Barillot E, Bluthgen N, Dekker J, Heard E (2012) Spatial partitioning of the regulatory landscape of the X-inactivation centre. Nature 485(7398):381–385. https://doi.org/10.1038/nature11049

Nora EP, Goloborodko A, Valton AL, Gibcus JH, Uebersohn A, Abdennur N, Dekker J, Mirny LA, Bruneau BG (2017) Targeted degradation of CTCF decouples local insulation of chromosome domains from genomic compartmentalization. Cell 169(5):930–944. e922. https://doi.org/10.1016/j.cell.2017.05.004

Nuebler J, Fudenberg G, Imakaev M, Abdennur N, Mirny LA (2018) Chromatin organization by an interplay of loop extrusion and compartmental segregation. Proc Natl Acad Sci U S A 115(29):E6697–E6706. https://doi.org/10.1073/pnas.1717730115

Pepenella S, Murphy KJ, Hayes JJ (2014) Intra- and inter-nucleosome interactions of the core histone tail domains in higher-order chromatin structure. Chromosoma 123(1-2):3–13. https://doi.org/10.1007/s00412-013-0435-8

Phillips-Cremins JE, Sauria ME, Sanyal A, Gerasimova TI, Lajoie BR, Bell JS, Ong CT, Hookway TA, Guo C, Sun Y, Bland MJ, Wagstaff W, Dalton S, McDevitt TC, Sen R, Dekker J, Taylor J, Corces VG (2013) Architectural protein subclasses shape 3D organization of genomes during lineage commitment. Cell 153(6):1281–1295. https://doi.org/10.1016/j.cell.2013.04.053

Philonenko ES, Klochkov DB, Borunova VV, Gavrilov AA, Razin SV, Iarovaia OV (2009) TMEM8 – a non-globin gene entrapped in the globin web. Nucleic Acids Res 37(22):7394–7406

Pickersgill H, Kalverda B, de Wit E, Talhout W, Fornerod M, van Steensel B (2006) Characterization of the Drosophila melanogaster genome at the nuclear lamina. Nat Genet 38(9):1005–1014. https://doi.org/10.1038/ng1852

Pliss A, Malyavantham KS, Bhattacharya S, Berezney R (2013) Chromatin dynamics in living cells: identification of oscillatory motion. J Cell Physiol 228(3):609–616. https://doi.org/10.1002/jcp.24169

Pombo A, Nicodemi M (2014) Physical mechanisms behind the large scale features of chromatin organization. Transcription 5(2):e28447. https://doi.org/10.4161/trns.28447

Pope BD, Ryba T, Dileep V, Yue F, Wu W, Denas O, Vera DL, Wang Y, Hansen RS, Canfield TK, Thurman RE, Cheng Y, Gulsoy G, Dennis JH, Snyder MP, Stamatoyannopoulos JA, Taylor J, Hardison RC, Kahveci T, Ren B, Gilbert DM (2014) Topologically associating domains are stable units of replication-timing regulation. Nature 515(7527):402–405. https://doi.org/10.1038/nature13986

Quinodoz SA, Ollikainen N, Tabak B, Palla A, Schmidt JM, Detmar E, Lai MM, Shishkin AA, Bhat P, Takei Y, Trinh V, Aznauryan E, Russell P, Cheng C, Jovanovic M, Chow A, Cai L, McDonel P, Garber M, Guttman M (2018) Higher-order inter-chromosomal hubs shape 3D genome organization in the nucleus. Cell 174(3):744–757. e724. https://doi.org/10.1016/j.cell.2018.05.024

Rada-Iglesias A, Grosveld FG, Papantonis A (2018) Forces driving the three-dimensional folding of eukaryotic genomes. Mol Syst Biol 14(6):e8214. https://doi.org/10.15252/msb.20188214

Rao SS, Huntley MH, Durand NC, Stamenova EK, Bochkov ID, Robinson JT, Sanborn AL, Machol I, Omer AD, Lander ES, Aiden EL (2014) A 3D map of the human genome at kilo-base resolution reveals principles of chromatin looping. Cell 159(7):1665–1680. https://doi.org/10.1016/j.cell.2014.11.021

Rao SSP, Huang SC, Glenn St Hilaire B, Engreitz JM, Perez EM, Kieffer-Kwon KR, Sanborn AL, Johnstone SE, Bascom GD, Bochkov ID, Huang X, Shamim MS, Shin J, Turner D, Ye Z, Omer AD, Robinson JT, Schlick T, Bernstein BE, Casellas R, Lander ES, Aiden EL (2017) Cohesin loss eliminates all loop domains. Cell 171(2):305–320. e324. https://doi.org/10.1016/j.cell.2017.09.026

Razin SV, Vassetzky YS (2017) 3D genomics imposes evolution of the domain model of eukaryotic genome organization. Chromosoma 126:59–69. https://doi.org/10.1007/s00412-016-0604-7

Razin SV, Gavrilov AA, Pichugin A, Lipinski M, Iarovaia OV, Vassetzky YS (2011) Transcription factories in the context of the nuclear and genome organization. Nucleic Acids Res 39(21):9085–9092. https://doi.org/10.1093/nar/gkr683

Razin SV, Gavrilov AA, Ioudinkova ES, Iarovaia OV (2013) Communication of genome regulatory elements in a folded chromosome. FEBS Lett 587(13):1840–1847. https://doi.org/10.1016/j.febslet.2013.04.027

Rippe K (2007) Dynamic organization of the cell nucleus. Curr Opin Genet Dev 17(5):373–380. https://doi.org/10.1016/j.gde.2007.08.007

Rosa A, Everaers R (2008) Structure and dynamics of interphase chromosomes. PLoS Comput Biol 4(8):e1000153. https://doi.org/10.1371/journal.pcbi.1000153

Rowley MJ, Corces VG (2018) Organizational principles of 3D genome architecture. Nat Rev Genet 19(12):789–800. https://doi.org/10.1038/s41576-018-0060-8

Rowley MJ, Nichols MH, Lyu X, Ando-Kuri M, Rivera ISM, Hermetz K, Wang P, Ruan Y, Corces VG (2017) Evolutionarily conserved principles predict 3D chromatin organization. Mol Cell. https://doi.org/10.1016/j.molcel.2017.07.022

Ruf S, Symmons O, Uslu VV, Dolle D, Hot C, Ettwiller L, Spitz F (2011) Large-scale analysis of the regulatory architecture of the mouse genome with a transposon-associated sensor. Nat Genet 43(4):379–386. https://doi.org/10.1038/ng.790

Sahlen P, Abdullayev I, Ramskold D, Matskova L, Rilakovic N, Lotstedt B, Albert TJ, Lundeberg J, Sandberg R (2015) Genome-wide mapping of promoter-anchored interactions with close to single-enhancer resolution. Genome Biol 16:156. https://doi.org/10.1186/s13059-015-0727-9

Sanborn AL, Rao SS, Huang SC, Durand NC, Huntley MH, Jewett AI, Bochkov ID, Chinnappan D, Cutkosky A, Li J, Geeting KP, Gnirke A, Melnikov A, McKenna D, Stamenova EK, Lander ES, Aiden EL (2015) Chromatin extrusion explains key features of loop and domain formation in wild-type and engineered genomes. Proc Natl Acad Sci U S A 112(47):E6456–E6465. https://doi.org/10.1073/pnas.1518552112

Schneider R, Grosschedl R (2007) Dynamics and interplay of nuclear architecture, genome organization, and gene expression. Genes Dev 21(23):3027–3043. https://doi.org/10.1101/gad.1604607

Schoenfelder S, Sexton T, Chakalova L, Cope NF, Horton A, Andrews S, Kurukuti S, Mitchell JA, Umlauf D, Dimitrova DS, Eskiw CH, Luo Y, Wei CL, Ruan Y, Bieker JJ, Fraser P (2010) Preferential associations between co-regulated genes reveal a transcriptional interactome in erythroid cells. Nat Genet 42(1):53–61

Sengupta K (2018) Genome 3D-architecture: its plasticity in relation to function. J Biosci 43(2):417–419

Sexton T, Yaffe E, Kenigsberg E, Bantignies F, Leblanc B, Hoichman M, Parrinello H, Tanay A, Cavalli G (2012) Three-dimensional folding and functional organization principles of the Drosophila genome. Cell 148(3):458–472. https://doi.org/10.1016/j.cell.2012.01.010

Shah FR, Bhat YA, Wani AH (2018) Subnuclear distribution of proteins: Links with genome architecture. Nucleus 9(1):42–55. https://doi.org/10.1080/19491034.2017.1361578

Shevelyov YY, Nurminsky DI (2012) The nuclear lamina as a gene-silencing hub. Curr Issues Mol Biol 14(1):27–38

Shogren-Knaak M, Ishii H, Sun JM, Pazin MJ, Davie JR, Peterson CL (2006) Histone H4-K16 acetylation controls chromatin structure and protein interactions. Science 311(5762):844–847. https://doi.org/10.1126/science.1124000

Shopland LS, Johnson CV, Byron M, McNeil J, Lawrence JB (2003) Clustering of multiple specific genes and gene-rich R-bands around SC-35 domains: evidence for local euchromatic neighborhoods. J Cell Biol 162(6):981–990. https://doi.org/10.1083/jcb.200303131

Sinha D, Shogren-Knaak MA (2010) Role of direct interactions between the histone H4 Tail and the H2A core in long range nucleosome contacts. J Biol Chem 285(22):16572–16581. https://doi.org/10.1074/jbc.M109.091298

Smeets D, Markaki Y, Schmid VJ, Kraus F, Tattermusch A, Cerase A, Sterr M, Fiedler S, Demmerle J, Popken J, Leonhardt H, Brockdorff N, Cremer T, Schermelleh L, Cremer M (2014) Three-dimensional super-resolution microscopy of the inactive X chromosome territory reveals a collapse of its active nuclear compartment harboring distinct Xist RNA foci. Epigenet Chromatin 7:8. https://doi.org/10.1186/1756-8935-7-8

Sofueva S, Yaffe E, Chan WC, Georgopoulou D, Vietri Rudan M, Mira-Bontenbal H, Pollard SM, Schroth GP, Tanay A, Hadjur S (2013) Cohesin-mediated interactions organize chromosomal domain architecture. EMBO J 32(24):3119–3129. https://doi.org/10.1038/emboj.2013.237

Spitz F, Gonzalez F, Duboule D (2003) A global control region defines a chromosomal regulatory landscape containing the HoxD cluster. Cell 113(3):405–417

Stanek D, Fox AH (2017) Nuclear bodies: news insights into structure and function. Curr Opin Cell Biol 46:94–101. https://doi.org/10.1016/j.ceb.2017.05.001

Stevens TJ, Lando D, Basu S, Atkinson LP, Cao Y, Lee SF, Leeb M, Wohlfahrt KJ, Boucher W, O'Shaughnessy-Kirwan A, Cramard J, Faure AJ, Ralser M, Blanco E, Morey L, Sanso M, Palayret MGS, Lehner B, Di Croce L, Wutz A, Hendrich B, Klenerman D, Laue ED (2017) 3D structures of individual mammalian genomes studied by single-cell Hi-C. Nature 544(7648):59–64. https://doi.org/10.1038/nature21429

Stigler J, Camdere GO, Koshland DE, Greene EC (2016) Single-molecule imaging reveals a collapsed conformational state for DNA-bound cohesin. Cell Rep 15(5):988–998. https://doi.org/10.1016/j.celrep.2016.04.003

Sun Y, Durrin LK, Krontiris TG (2003) Specific interaction of PML bodies with the TP53 locus in Jurkat interphase nuclei. Genomics 82(2):250–252

Sun F, Chronis C, Kronenberg M, Chen XF, Su T, Lay FD, Plath K, Kurdistani SK, Carey MF (2019) Promoter-enhancer communication occurs primarily within insulated neighborhoods. Mol Cell 73(2):250–263. e255. https://doi.org/10.1016/j.molcel.2018.10.039

Symmons O, Uslu VV, Tsujimura T, Ruf S, Nassari S, Schwarzer W, Ettwiller L, Spitz F (2014) Functional and topological characteristics of mammalian regulatory domains. Genome Res 24(3):390–400. https://doi.org/10.1101/gr.163519.113

Symmons O, Pan L, Remeseiro S, Aktas T, Klein F, Huber W, Spitz F (2016) The Shh topological domain facilitates the action of remote enhancers by reducing the effects of genomic distances. Dev Cell 39(5):529–543. https://doi.org/10.1016/j.devcel.2016.10.015

Szabo Q, Jost D, Chang JM, Cattoni DI, Papadopoulos GL, Bonev B, Sexton T, Gurgo J, Jacquier C, Nollmann M, Bantignies F, Cavalli G (2018) TADs are 3D structural units of higher-order chromosome organization in Drosophila. Sci Adv 4(2):eaar8082. https://doi.org/10.1126/sciadv.aar8082

Szczerbal I, Bridger JM (2010) Association of adipogenic genes with SC-35 domains during porcine adipogenesis. Chromosome Res 18(8):887–895. https://doi.org/10.1007/s10577-010-9176-1

Tark-Dame M, van Driel R, Heermann DW (2011) Chromatin folding--from biology to polymer models and back. J Cell Sci 124(Pt 6):839–845. https://doi.org/10.1242/jcs.077628

Tiana G, Amitai A, Pollex T, Piolot T, Holcman D, Heard E, Giorgetti L (2016) Structural fluctuations of the chromatin fiber within topologically associating domains. Biophys J 110(6):1234–1245. https://doi.org/10.1016/j.bpj.2016.02.003

Tolhuis B, Palstra RJ, Splinter E, Grosveld F, de Laat W (2002) Looping and interaction between hypersensitive sites in the active beta-globin locus. Mol Cell 10(6):1453–1465. https://doi. org/10.1016/S1097-2765(02)00781-5

Udvardy A, Schedl P, Sander M, Hsieh T-S (1986) J Mol Biol 191:231–246

Ulianov SV, Gavrilov AA, Razin SV (2015) Nuclear compartments, genome folding, and enhancer-promoter communication. Int Rev Cell Mol Biol 315:183–244. https://doi.org/10.1016/ bs.ircmb.2014.11.004

Ulianov SV, Khrameeva EE, Gavrilov AA, Flyamer IM, Kos P, Mikhaleva EA, Penin AA, Logacheva MD, Imakaev MV, Chertovich A, Gelfand MS, Shevelyov YY, Razin SV (2016) Active chromatin and transcription play a key role in chromosome partitioning into topologically associating domains. Genome Res 26(1):70–84. https://doi.org/10.1101/gr.196006.115

Ulianov SV, Galitsyna AA, Flyamer IM, Golov AK, Khrameeva EE, Imakaev MV, Abdennur NA, Gelfand MS, Gavrilov AA, Razin SV (2017) Activation of the alpha-globin gene expression correlates with dramatic upregulation of nearby non-globin genes and changes in local and large-scale chromatin spatial structure. Epigenet Chromatin 10(1):35. https://doi.org/10.1186/ s13072-017-0142-4

Valton AL, Dekker J (2016) TAD disruption as oncogenic driver. Curr Opin Genet Dev 36:34–40. https://doi.org/10.1016/j.gde.2016.03.008

van de Werken HJ, Landan G, Holwerda SJ, Hoichman M, Klous P, Chachik R, Splinter E, Valdes-Quezada C, Oz Y, Bouwman BA, Verstegen MJ, de Wit E, Tanay A, de Laat W (2012) Robust 4C-seq data analysis to screen for regulatory DNA interactions. Nat Methods 9(10):969–972. https://doi.org/10.1038/nmeth.2173

van Koningsbruggen S, Gierlinski M, Schofield P, Martin D, Barton GJ, Ariyurek Y, den Dunnen JT, Lamond AI (2010) High-resolution whole-genome sequencing reveals that specific chromatin domains from most human chromosomes associate with nucleoli. Mol Biol Cell 21(21):3735–3748. https://doi.org/10.1091/mbc.E10-06-0508

van Steensel B, Belmont AS (2017) Lamina-associated domains: links with chromosome architecture, heterochromatin, and gene repression. Cell 169(5):780–791. https://doi.org/10.1016/j. cell.2017.04.022

Vasquez PA, Hult C, Adalsteinsson D, Lawrimore J, Forest MG, Bloom K (2016) Entropy gives rise to topologically associating domains. Nucleic Acids Res 44(12):5540–5549. https://doi. org/10.1093/nar/gkw510

Vernimmen D, Bickmore WA (2015) The hierarchy of transcriptional activation: from enhancer to promoter. Trends Genet 31(12):696–708. https://doi.org/10.1016/j.tig.2015.10.004

Vernimmen D, De Gobbi M, Sloane-Stanley JA, Wood WG, Higgs DR (2007) Long-range chromosomal interactions regulate the timing of the transition between poised and active gene expression. EMBO J 26(8):2041–2051. https://doi.org/10.1038/sj.emboj.7601654

Vernimmen D, Marques-Kranc F, Sharpe JA, Sloane-Stanley JA, Wood WG, Wallace HA, Smith AJ, Higgs DR (2009) Chromosome looping at the human alpha-globin locus is mediated via the major upstream regulatory element (HS -40). Blood 114(19):4253–4260. https://doi. org/10.1182/blood-2009-03-213439

Vian L, Pekowska A, Rao SSP, Kieffer-Kwon KR, Jung S, Baranello L, Huang SC, El Khattabi L, Dose M, Pruett N, Sanborn AL, Canela A, Maman Y, Oksanen A, Resch W, Li X, Lee B, Kovalchuk AL, Tang Z, Nelson S, Di Pierro M, Cheng RR, Machol I, St Hilaire BG, Durand NC, Shamim MS, Stamenova EK, Onuchic JN, Ruan Y, Nussenzweig A, Levens D, Aiden EL, Casellas R (2018) The energetics and physiological impact of cohesin extrusion. Cell 173(5):1165–1178. e1120. https://doi.org/10.1016/j.cell.2018.03.072

Vicente-Garcia C, Villarejo-Balcells B, Irastorza-Azcarate I, Naranjo S, Acemel RD, Tena JJ, Rigby PWJ, Devos DP, Gomez-Skarmeta JL, Carvajal JJ (2017) Regulatory landscape fusion in rhabdomyosarcoma through interactions between the PAX3 promoter and FOXO1 regulatory elements. Genome Biol 18(1):106. https://doi.org/10.1186/s13059-017-1225-z

Vietri Rudan M, Hadjur S (2015) Genetic tailors: CTCF and cohesin shape the genome during evolution. Trends Genet 31(11):651–660. https://doi.org/10.1016/j.tig.2015.09.004

Vietri Rudan M, Barrington C, Henderson S, Ernst C, Odom DT, Tanay A, Hadjur S (2015) Comparative Hi-C reveals that CTCF underlies evolution of chromosomal domain architecture. Cell Rep 10(8):1297–1309. https://doi.org/10.1016/j.celrep.2015.02.004

Wang G, Achim CL, Hamilton RL, Wiley CA, Soontornniyomkij V (1999) Tyramide signal amplification method in multiple-label immunofluorescence confocal microscopy. Methods 18(4):459–464. https://doi.org/10.1006/meth.1999.0813

Wang J, Shiels C, Sasieni P, Wu PJ, Islam SA, Freemont PS, Sheer D (2004) Promyelocytic leukemia nuclear bodies associate with transcriptionally active genomic regions. J Cell Biol 164(4):515–526. https://doi.org/10.1083/jcb.200305142

Wang C, Liu C, Roqueiro D, Grimm D, Schwab R, Becker C, Lanz C, Weigel D (2015) Genome-wide analysis of local chromatin packing in Arabidopsis thaliana. Genome Res 25(2):246–256. https://doi.org/10.1101/gr.170332.113

Wang Q, Sawyer IA, Sung MH, Sturgill D, Shevtsov SP, Pegoraro G, Hakim O, Baek S, Hager GL, Dundr M (2016) Cajal bodies are linked to genome conformation. Nat Commun 7:10966. https://doi.org/10.1038/ncomms10966

Weinreb C, Raphael BJ (2016) Identification of hierarchical chromatin domains. Bioinformatics 32(11):1601–1609. https://doi.org/10.1093/bioinformatics/btv485

Weintraub H, Groudine M (1976) Chromosomal subunits in active genes have an altered conformation. Science 73:848–856

Weintraub H, Larsen A, Groudine M (1981) Alpha-Globin-gene switching during the development of chicken embryos: expression and chromosome structure. Cell 24(2):333–344

West AG, Fraser P (2005) Remote control of gene transcription. Hum Mol Genet 14 Spec No 1:R101–R111

Wutz G, Varnai C, Nagasaka K, Cisneros DA, Stocsits RR, Tang W, Schoenfelder S, Jessberger G, Muhar M, Hossain MJ, Walther N, Koch B, Kueblbeck M, Ellenberg J, Zuber J, Fraser P, Peters JM (2017) Topologically associating domains and chromatin loops depend on cohesin and are regulated by CTCF, WAPL, and PDS5 proteins. EMBO J 36(24):3573–3599. https://doi.org/10.15252/embj.201798004

Zhan Y, Mariani L, Barozzi I, Schulz EG, Bluthgen N, Stadler M, Tiana G, Giorgetti L (2017) Reciprocal insulation analysis of Hi-C data shows that TADs represent a functionally but not structurally privileged scale in the hierarchical folding of chromosomes. Genome Res 27(3):479–490. https://doi.org/10.1101/gr.212803.116

Zirbel RM, Mathieu UR, Kurz A, Cremer T, Lichter P (1993) Evidence for a nuclear compartment of transcription and splicing located at chromosome domain boundaries. Chromosome Res 1(2):93–106

Zolotarev N, Fedotova A, Kyrchanova O, Bonchuk A, Penin AA, Lando AS, Eliseeva IA, Kulakovskiy IV, Maksimenko O, Georgiev P (2016) Architectural proteins Pita, Zw5, and ZIPIC contain homodimerization domain and support specific long-range interactions in Drosophila. Nucleic Acids Res 44(15):7228–7241. https://doi.org/10.1093/nar/gkw371

Zuin J, Dixon JR, van der Reijden MI, Ye Z, Kolovos P, Brouwer RW, van de Corput MP, van de Werken HJ, Knoch TA, van IWF, Grosveld FG, Ren B, Wendt KS (2014) Cohesin and CTCF differentially affect chromatin architecture and gene expression in human cells. Proc Natl Acad Sci U S A 111(3):996–1001. https://doi.org/10.1073/pnas.1317788111

Zuniga A, Michos O, Spitz F, Haramis AP, Panman L, Galli A, Vintersten K, Klasen C, Mansfield W, Kuc S, Duboule D, Dono R, Zeller R (2004) Mouse limb deformity mutations disrupt a global control region within the large regulatory landscape required for Gremlin expression. Genes Dev 18(13):1553–1564. https://doi.org/10.1101/gad.299904

Chapter 3
Analysis of Cell and Nucleus Genome by Next-Generation Sequencing

Ji Won Oh and Alexej Abyzov

Abstract Genomic variants that are acquired during a lifetime as a result of development, environmental exposure, and aging are present in every cell of the human body. While some variants are shared between cells, most of them are not. Therefore, analyzing the nuclear genome of a cell is the ultimate way to study genomic mosaicism. However, comprehensive evaluation of variations in a single cell's genome is not yet possible due to unresolved technical issues, while the analysis of a bulk of cells can provide a valuable insight into mosaicism in a studied sample. Here, we describe, compare, and discuss strategies, experimental techniques, and analytical methods for discovery of a spectrum of mosaic variants from a bulk of cells and from single cells. We specifically focus on next-generation sequencing technologies for genome analysis as they enable the discovery of mosaic variants of all types.

Mosaic Variants During Lifetime

Introduction

Mosaic genome variation is a difference in the DNA sequence between cells of the same individual. Such differences can be as small as one nucleotide and as large as the entire chromosome and are subdivided into the following types: single nucleotide variants (SNV), small insertions and deletions (indel), copy number alterations (CNA), copy number neutral losses of heterozygosity (CNN-LOH), sequence

J. W. Oh
Department of Anatomy, School of Medicine, Kyungpook National University,
Daegu, South Korea

Biomedical Research Institute, Kyungpook National University Hospital, Daegu, South Korea

A. Abyzov (✉)
Department of Health Sciences Research, Center for Individualized Medicine, Mayo Clinic,
Rochester, MN, USA
e-mail: abyzov.alexej@mayo.edu

© Springer Nature Switzerland AG 2020
I. Iourov et al. (eds.), *Human Interphase Chromosomes*,
https://doi.org/10.1007/978-3-030-62532-0_3

35

Box 3.1 Types of mosaic genomic variations

Single nucleotide variant (**SNV**) is a difference at a particular genomic position in a single nucleotide such as A to G.

Indel is an insertion (in-) or a deletion (−del) of a few consecutive nucleotides with typical size of less than 50 bps or 100 bps.

Copy number alteration (**CNA**) is a region with fewer (deletion) or more (duplication) copies of DNA. CNAs typically refer to somatic alterations and are distinguished from indels by size (CNAs are larger). But mechanistic origin of indels and CNA could also be different.

Copy number neutral loss of heterozygosity (**CNN-LOH**) is a genomic region where two haplotypes (paternal and maternal) are identical. The term is used to describe acquired mutations.

Inversion is a replacement of a 5′–3′ nucleotide sequence end to end with the reverse complement.

Interchromosomal translocation is the rearrangement that connects sequences of two different chromosomes.

Chromosomal aneuploidy is the variation in the number of copies for a chromosome.

Multiploidy is global change in the number of chromosomes.

Structural variant (**SV**) is a general category that encompasses CNAs, CNN-LOH, inversions, interchromosomal translocations, chromosomal aneuploidies, multiploidies, MEIs, and complex rearrangements that carry signatures of multiple just-listed types.

Mobile element insertion (**MEI**) is an insertion into the genome of a retrotransposon. In the human genome, four retrotransposon families are known, ALU, LINE1, SVA, and HERV, but the latter is believed to be inactive (Solyom and Kazazian 2012).

inversions, interchromosomal translocations, chromosomal aneuploidies, multiploidies, and insertions of retrotransposable elements (Box 3.1). For generalization, variants other than SNVs and indels are called structural variants (SVs). Mosaic variants of all types have commonalities relevant to their detection. Variants with low frequency in cells of an individual are harder to discover as their contribution to the experimental measurable signal used for discovery (e.g., sequencing data) is also small. The rarity of variants can be the result of them being disadvantageous to the cells carrying them. For instance, variants that reduce cell proliferation or viability are likely to be less frequent in a tissue and thus are harder to detect experimentally. At an extreme, mosaic variants leading to cell death for whatever reason cannot be discovered at all. On the other hand, variants that increase proliferation or viability will likely be more frequent (Poduri et al. 2012) and, consequently, are easier to detect. Certain variants could interfere with the experimental procedure for data generation and variant detection such as cell cloning or transformation, preventing the discovery of the variants in such conditions.

Mosaic Variants in Development and Aging

Right after conception, hundreds of spontaneous mutations begin to accumulate in cells of a developing fetus, a process that continues at a much slower rate well into adulthood. Every time a cell divides, its two progenies have shorter telomeres and are likely to acquire mutations from replication errors (Kunkel 2004; Aubert and Lansdorp 2008). These variants are then passed on to their progenies at the next divisions. Additionally, unrepaired or incorrectly repaired DNA damage also results in mutations. The damage can be either spontaneous, for example, deamination caused by interaction with reactive oxygen species (Bacolla et al. 2014), or environmentally driven, for example, caused by ultraviolet radiation from the sun exposure (Ikehata and Ono 2011). It is therefore very likely that in every human being, there are no two single cells with identical genomes. Moreover, as variations in each cell accumulate with time (Ramsey et al. 1995; Jacobs et al. 2012; Laurie et al. 2012; Forsberg et al. 2012; Blokzijl et al. 2016; Lodato et al. 2018), cellular genomes in every individual diverge, i.e., become more different from each other.

However, the rates at which genomes accumulate mutations in development and during aging are only partially known. Our ability to measure mutation rates in development is fundamentally limited by the common ethics prohibiting experiments with human embryos. The fundamental challenge for the estimation of mutation rates in the context of aging is the rarity of mutations accumulated by each cell. Namely, variants acquired by a cell in adulthood are quite likely to exist only in that single cell, making their discovery technologically challenging and experimental validation impossible in principle. The ability to comprehensively characterize the entire genome of a single cell is lacking. Most experimental techniques require single-cell genome amplification, which, as discussed below, is a technical challenge, while a few that do not, like fluorescent in situ hybridization, provide only crude resolution. A notable exception to this challenge is stem cells of every tissue and organ. These cells will pass acquired mutations to their offsprings (Blokzijl et al. 2016; Bae et al. 2018) and are also culturable, allowing to bypass error-prone in vitro amplification.

Initially, mutation rates were estimated indirectly. Based on the frequency (50 to 100) of de novo SNVs in genomes of newborn humans, it was estimated that cells in the germline lineage acquired 1.2 SNVs per cell division during development and about 0.1–0.2 SNVs per cell division post puberty (Conrad et al. 2010; Michaelson et al. 2012; Rahbari et al. 2016; Maretty et al. 2017; Yuen et al. 2017; Milholland et al. 2017). Recent studies provided a more direct insight into the rates. Reconstruction of a cell progeny tree during cleavages yielded the rate of 1.3 SNVs per division per progeny (Bae et al. 2018). As such cells contribute to all germ lineages including germline, this estimate is consistent with the one provided by studies of de novo SNVs. The same study suggested a higher mutation rate during neurogenesis and a shift in mutation spectrum from dominantly C > T to dominantly C > A mutations (Bae et al. 2018). Similar results were obtained by another group (Kuijk et al. 2019). Overall higher mutation rates during organogenesis is consistent with the results of measuring mutation load across several tissues (Blokzijl et al. 2016; Lodato et al. 2018; Franco et al. 2018; Abyzov et al. 2017). It was revealed that already at birth, muscle, neurons, skin fibroblasts, intestine, colon, and liver cells have a similar order of 500–1000 mutations.

The same studies have also estimated that the rate of accumulation of mosaic SNVs postnatally is rather slow, from 10 to 60 SNVs per year. Additionally, a gain in knowledge about genome variation with age has been obtained from indirect studies, such as analyses of cancer genomes. As almost all cancers are expanded from a single cell, the genome of the founder cell is replicated in every cell of a cancer. However, variant frequency in a given cancer sample is not a good indication of time of origin because secondary mutations that occur in cancer cells can rise to high frequency due to subclonal expansion. Apart from that, cancer samples are often not perfectly clonal because of the admixture of normal and immune cells. Despite all that, multiple lines of evidence suggest that most mutations observed in cancer originate in healthy cells prior to malignant transformation, and their analysis is informative about mutagenesis in normal cells and its relationship to aging (Tomasetti et al. 2013; Milholland et al. 2015; Lee et al. 2019). A rather simple cross correspondence of mutation burden in cancers with patient age estimated a somatic mutation rate of about 100 SNVs per cell per year in over 20 tissues (Milholland et al. 2017; Podolskiy et al. 2016).

A better estimate is possible by decomposing the spectrum of SNVs across thousands of cancers of different types into individual components, the so-called mutation signatures (Alexandrov et al. 2013; Lawrence et al. 2013). Some of these signatures match to known mutational mechanisms such as damage by ultraviolet light and DNA editing by the APOBEC enzyme. The overall thought is that most of these signatures represent mutational processes. Together with the decomposition, it is also possible to estimate the contribution of each signature to the mutation spectrum of each individual cancer. Consequently, one can analyze the contribution of each signature in relation to an individual's age. It was found that only two signatures, #1 and #5, showed a monotonic increase with age in most cancer types, suggesting that they represent mutational processes during natural cell aging (Alexandrov et al. 2015). By considering only those signatures, the mutation rate across various tissues in adults was estimated to vary from a few to 30 SNVs per genome per year (Alexandrov et al. 2015). These estimates are consistent with the direct measurements of mutation accumulation in the liver, colon, and small intestine (Blokzijl et al. 2016). A similar mutation rate in neurons was reported by a single-cell study (Lodato et al. 2018). At the same time, it is also clear that the environment can have a dramatic impact on mutation burden in a cell. Specifically, ultraviolet light damage can result in an order of magnitude higher mutation count seen in a cell (Saini et al. 2016).

Contrary to somatic tissues, germline lineage was consistently observed to have lower background mutability. Regression on parental age implies that variants detected as de novo in children accumulate at a rate of two to three SNVs per year in father sperm and 0.5–1 SNVs in mother oocytes (Michaelson et al. 2012; Rahbari et al. 2016; Maretty et al. 2017), i.e., about an order of magnitude lower mutability. Other evidence suggests an even larger difference of two orders of magnitude (Milholland et al. 2017).

Less insight was gained into the association with the aging of variant types other than SNVs. This was dictated by their – variants of other types – overall less-frequent occurrence, difficulties in their discovery, and challenges in deriving proper descriptors for the mutational spectrum. Somatic CNAs were observed in at least seven human tissues (O'Huallachain et al. 2012). Analysis of SNP arrays for DNA from blood from over 50,000 individuals revealed that CNAs larger than 2 mbp are found in less than 0.5% of individuals younger than 50 years but are substantially more frequent (in about 2%) in elderly people (older than 70 years) (Jacobs et al. 2012; Laurie et al. 2012). Recently, a more sensitive (toward lower-frequency CNAs) study using improved analytical methodology updated those estimates, showing that rare CNAs are detectable in 5% of people (Loh et al. 2018). The major limitation of these studies is that they are based on SNP arrays and can only detect large CNAs. They, therefore, only inform about the most frequent and, likely, expanded cell clones.

A direct estimation of how many cells have CNAs and how their frequency and spectrum change during a lifetime was possible from single-cell studies. It was estimated that about 30% of fibroblast cells and neuronal nuclei carry CNAs (Abyzov et al. 2012; McConnell et al. 2013; Cai et al. 2014). It was, however, suggested that CNAs in neurons decrease with age (Chronister et al. 2019). Aneuploidies were detected in liver hepatocytes, oocytes, and neurons (Duncan et al. 2012; Jones 2008; Yurov et al. 2007). Their frequency, however, is debated (Knouse et al. 2014). Additionally, it is known that somatic insertions of mobile elements such as L1 are present in brain cells; however, similar to aneuploidies, their frequency is controversial, and their association with age is unknown (Evrony et al. 2012; Evrony et al. 2015; Baillie et al. 2011; Erwin et al. 2016; Fig. 3.1).

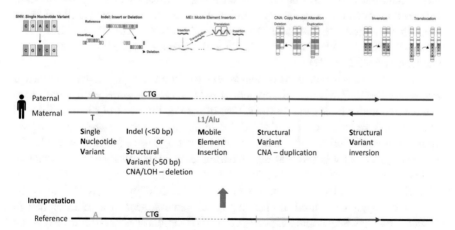

Fig. 3.1 Depiction of main variant types in the human genome (see Box 1 for definitions). For germline variants, the type of variant call may depend on the sequence of the reference genome (lower panel). For example, mobile element insertion, if present in the reference, will be discovered as deletion. For mosaic variants, there is no such ambiguity. LOH stands for loss of heterozygosity

Application of Mosaic Variants for Cell Lineage Tracing

In 1892, Edmund Wilson used "cell lineage" when he analyzed the contribution to the cytogeny of the annelid body (Wilson 1892). He found that each cell has the trajectory of continuous divisional events based on their fates, leading to the concept of "cell lineage." Thirteen years after the work of E. Wilson, in 1905, E.G. Conklin expanded the concept by defining invariant and non-invariant cell lineages based on the investigation of ascidian egg (Conklin 1905). E. Wilson and E.G. Conklin used their bare eyes under the optical microscope without any molecular intervention or chemical staining of the cells. Their works needed to investigate each cell very carefully, and they had to infer the divisional trajectory based on topological contribution, limiting their speculation only to the early cellular division of embryological cleavages. It was almost impossible to track the cells after certain divisions due to the relatively large cell number, necessitating a new method. In 1929, Vogt utilized the chemical staining for the tracking of groups of cells (Vogt 1929). His work with the grafting experiment of Spemann and Mangold expanded the concept of "cell lineage" to the amphibian germ cell study (Spemann and Evolution 1924). The chemical stain could give clear trajectory evidence on cellular division, although the stain was diluted after each division.

Intriguingly, in the same year that Spemann and Mangold used amphibian tissues for tracking cellular lineages, Sturtevant used the large-scale structural variants, chromosomal elimination, for the cellular lineage tracking in Drosophila (Sturtevant 1929). Before this work, the group of T.H. Morgan developed the experimental methods to generate spontaneous mosaicism, which is inherited by divisional daughter cell. Sturtevant investigated the cell of the insects retrospectively and found that their chromosomal elimination is highly linked with the gene function. His work inspired not only the distribution of ring X chromosomal study but many other lineage tracking studies of Drosophila (Catcheside et al. 1945; Wald 1936; Garcia-Bellido and Merriam 1969; Hotta and Benzer 1973; Zalokar 1976; Ferrús and Garcia-Bellido 1977).

Before the development of a genetically modified animal model for lineage tracing in 1993 (Harrison and Perrimon 1993), microscopic inspection with bare eye led to a major biological findings. One of them is the work of Sir John Edward Sulston, who tracked every cellular division in *Caenorhabditis elegans*, showing the entire embryonic cell lineage tree of an individual organism (Sulston et al. 1983). His study eventually led to find the physiological cellular disappearance while an embryo keeps dividing, earning him the 2002 Nobel Prize for Physiology or Medicine for the discovery of programmed cell death, apoptosis. Additionally, he and his colleagues reported the first complete genome sequence of an animal in 1998 (C. elegans Sequencing Consortium 1998). Cellular lineage studies led to the development of key concepts in biology, not only apoptosis but cell commitment (Conklin 1905), cell fate potential (Tam et al. 1997), and cellular behaviors (Garcia-Bellido A. Cell Lineages and Genes 1985; Garcia-Bellido et al. 1973).

Lineage Tracing Using the Genetic Tools

To overcome the dilutional limitation of experimental methods using chemical stain, it was necessary to develop prospective genetic tools. Advances in recombinant DNA technology finally led to reporter genes transfer systems introduced into the live animal. Using retrovirus, they could incorporate the external gene sequences such as β-galactosidase and green fluorescent protein into the animal host genome (Turner and Cepko 1987; Frank and Sanes 1991). The integrated gene sequences are then inherited by the daughter cells of the infected founder cell. In principle, the approache works the same way as chemical stain injection experiments, however, since the reporter genes are genetically incorporated, there are no dilutional issues even though the cells keep dividing. Though retroviral labeling had some advantages over previous lineage tracing methods, there are several limitations. One of the limitations of using the retroviral gene transfer is that they can be incorporated into multiple cells at the first trial, making several founder cells. Researchers overcame this issue by limiting dilution assay of virus to label only one single founder cell. The other issue is the retroviral silence, leading to biased experimental results due to the under-evaluation of the number of descendants. If retroviral silencing happens, the transfer genes are not detected experimentally although their genome has the incorporated gene by retrovirus (Yao et al. 2004).

Tissue-specific genetic recombination tools broaden the fields of cellular lineage tracing to elucidate the various biological questions from the embryologic development to the adult stem cell biology. Genetic induction using FLP-FRT or Cre-loxP can control the incorporation of external DNA sequences spatially as well as temporally. In 1993, D.A. Harrison and N. Perrimon applied the site-specific yeast FLP recombinase combined with a heat shock-inducible promoter (Harrison and Perrimon 1993). The system in Drosophila successfully catalyzed the genetic recombination at the FRT (FLP recombination target), resulting in site-specific recombination inherited to every progeny from founder clone. In 1998, the Engrailed-Cre gene with b-actin-loxSTOPlox-lacZ gene was introduced into the mouse whole genome. Since the Engrailed gene was only expressed site-specifically, the lacZ turns on where Engrailed genes were expressed due to the Cre enzyme cleavage of loxP STOP codon sites (Zinyk et al. 1998). This experimental strategy elucidated the fate mapping of mouse midbrain-hindbrain constriction.

After these experiments, several tissue-specific promoters driving Cre systems were employed to elucidate the cellular behavior of stem cells (Snippert et al. 2010). Temporal control systems such as antibiotic-inducible systems (tetracycline-controlled transcriptional activation, Tet-On/Tet-Off) and hormone-dependent systems (tamoxifen-inducible CreER recombinase) were introduced as well (Gossen and Bujard 1992; Feil et al. 1997). The Brainbow system in neuron and Confetti system in the entire mouse body were reported to overcome, marking a limited number of stem cell lineages (Snippert et al. 2010; Livet et al. 2007). These systems utilized the stochastic selection of recombinase to choose multiple copies of fluorescent markers. After selection, each cell had unchanged different fluorescent color resulting from the random recombinase-mediated reporter sequences.

Theoretically, the system can distinguish around 90 cellular lineages depending on the possible combinations. The recent advent of intravital imaging technology combined with Cre-LoxP genetic recombination provides the window for in vivo stem cell behaviors (Yaniv et al. 2006).

Lineage Tracing Using Mosaic Variants

Due to the ethical reasons, we cannot incorporate the genes into the human or transfer extrinsic barcodes for the prospective lineage tracking. Thus, naturally occurring somatic mutation is an indispensable marker for the retrospective human lineage tracing experiment. In the case of genome replication including human species, there are several repair machineries for the accurate duplication of DNA (Friedberg 2003). However, due to the limited function of DNA polymerase as well as proofread machineries, de novo mosaic mutations occur in every cellular division. It is assumed that these mutations can happen inevitably without any functional cause, and therefore, their occurrence does not impair development. Somatic mutations occur stochastically all over the entire genome, making whole-genome sequencing (WGS) a requisite as an experimental tool in lineage tracing of human (De 2011). In a lineage, progenies can have the same shared mutations, which are inherited from the same ancestor, and certain accumulated somatic variants can be utilized for inferring the phylogenetic trajectory origin (Behjati et al. 2014; Shapiro et al. 2013). Recent studies are focusing on the early postzygotic variants as a form of somatic mutation for the lineage tracing (Bae et al. 2018; Behjati et al. 2014; Lodato et al. 2015), and such variants are shared by a large number of cells within an individual, so scientists can avoid some limitations of WGS errors bioinformatically and validate the true signal using a limited number of samples within a person.

In order to reconstruct a precise cellular lineage, strategies to avoid the false-positive errors from the biological experiments and to find the shared somatic variants across cells are important bioinformatically as well as experimentally. Because of the limitation of the substantial error rate of the current sequencing methods, there are various experimental strategies for the identification of somatic variants, such as single-cell whole-genome DNA sequencing, bulk sequencing, and in vitro clonal expansions. Each has its own advantages and limitations; we will discuss these strategies in detail. A critical reason for the development of each different strategy is mainly because experimental errors (false-positive signal) are more frequent than the genuine somatic variants.

In 2014, the first reconstruction of early developmental lineage using somatic variants was reported in mice (Behjati et al. 2014). Combined with the advances in sequencing technology and organoid cultivation methods, the authors observed an asymmetric contribution of the first reconstructed cell division, hypothesized by the previous developmental studies (Plusa et al. 2005; Bruce and Zernicka-Goetz 2010). The inequal contribution was repetitively confirmed in humans by several independent groups using different organs and experimental strategies (Ju et al. 2017; Lee-Six et al. 2018; Huang et al. 2018). Organoid culture techniques were

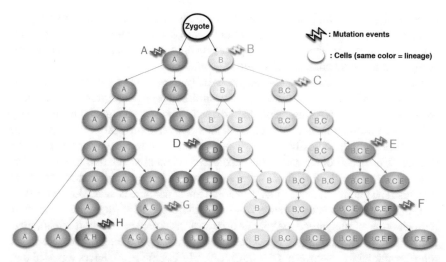

Fig. 3.2 Example of sequential events of early postzygotic somatic variants. When the zygote divides into two daughter cells, if the mutation "A" happens in the one cell and "B" in the other cell, one of the "A" or "B" mutations should be found in every cell of the analyzed sample. Furthermore, the shared mutations (including "A", "B" and later "C", "D", "E", etc.) can be utilized for the lineage reconstruction. Later arising mutation "F" and "G" will be shared by a relatively small number of cells. Rare or private mutations, such as "H," can only be found in one cell and are unlikely to be detected in bulk sequencing. The important experimental strategy to reconstruct cellular lineages is to segregate the shared mutations

especially useful to expand the single clones in vitro specifically in the internal organs; these were utilized to estimate the tissue-specific mutation rate of the human adult stem cell (Blokzijl et al. 2016). For the reconstruction of the early embryonic lineage, the critical analytical strategy in the clonal expansion method is to find the key somatic variants that are shared in at least two expansional clones as well as absent from at least one clone (Fig. 3.2).

Discovering Mosaic Variants from Next-Generation Sequencing

Next-generation sequencing (NGS) refers to multiple technologies enabling parallel (and, because of that, cost-effective) sequencing of fragmented DNA. A special preparation of fragmented DNA – the so-called library – is typically required to conduct the sequencing. From a prepared library, sequencing instruments output millions and even billions of reads – nucleotide sequences. To discover variants, the reads are first aligned against the reference genome in a procedure called mapping. Then the analysis of imperfections in the alignments allows one to discover variants in the studied genome relative to the human reference genome. For example, consistent mismatches or gaps in reads at a certain genomic position will indicate, respectively, SNVs and indels (Fig. 3.1). Comparison of two samples allows for the discovery of variants that are present in only one of the samples. Such comparison is most commonly made by comparing read mappings to the reference genome and rarely by direct read comparison of the two samples.

Currently, NGS technologies can be broadly classified as generating short or long reads (Table 3.1). Short reads come in pairs, and each pair represents sequences of the ends of a DNA fragment. Paired reads improve the mapping and carry additional information that can be used to discover structural variations in the studied genome. Specifically, paired reads are generated to have a certain expected distance from each other and certain expected orientation relative to each other. Loci with systematic deviation from such expectations likely harbor structural variants. The main advantage of short reads is lower per base cost and low error rate.

Long reads represent sequence from either end of DNA fragments. Due to at least an order of magnitude higher per base cost (as compared to short reads), their use is still not very common. However, because of the read length, they are much more beneficial for discovering SVs. Just recently, Pacific Biosciences reported that multiple readouts of the same read can provide superior sequencing quality while still keeping reads significantly long, on average 13.5 kbps (Wenger et al. 2019). This advance yet comes at significantly increased per base sequencing cost.

Overall, current sequencing technologies enable discovery across most of the human genome of all types of mosaic variants, and because of that, we will focus on describing their application to mosaic variant discovery. The efficiency of applying NGS to variant discovery depends on three key parameters: (i) read length, (ii) precision of sequencing, and (iii) depth of coverage. The longer the reads are, the better one can detect variants in repeats and avoid false positives. For short read technologies, the resolution of repeats depends both on read length and DNA fragment size. Reads come in pairs because they represent ends of the same DNA fragment and thus are considered as one entity for alignment and analysis. So effectively, the sequenced length is double the read length but not longer than the fragment size. But even with such consideration, short read technologies generate reads that are at least an order of magnitude shorter than those made by long read technologies. Regarding the other two parameters, the more the precise reads and the

Table 3.1 Characteristics of sequencing technologies

Characteristic	Short read	Long read
Companies that offer sequencing technologies	Illumina and BGI	Pacific Biosciences and Oxford Nanopore Technologies
Typical read length	100–150 bp	Over 20 kbp
Typical fragment length	450 bp	Unlimited
Reads are in pairs	Yes	No
Maximum read length	300 bps	Over 1 mbp
Sequencing error rate	~1% mostly mismatches	~10% mostly indels
Price per base	Lower	Higher

higher the coverage (both are a strength of short read technologies), the better and the more efficient is the discovery of variants.

Mosaic variants can be detected from genome analysis of a bulk of cells and analysis of single cells or nuclei (Fig. 3.3). In the first case, the genomes of many cells from brain or other tissue are investigated per experiment (Poduri et al. 2012). Analysis of bulk is frequently utilized in research studies because of easier, faster, and cheaper sample preparation/handling and data generation than those for single-cell analyses. One can extract and sequence DNA from all cells in a tissue/organ or from some cell fraction positive for a certain cell-type marker, thereby allowing the study of mosaicism in and between cell types (Matevossian and Akbarian 2008). Variant validation in the primary sample is relatively straightforward and can be done with more sensitive and orthogonal techniques than those used for discovery.

Sequencing data from bulk, however, represents a genome of at least a few hundred and more typically a few thousand cells, so the number of reads supporting a mosaic variant is proportional to its frequency in the studied sample. This is the

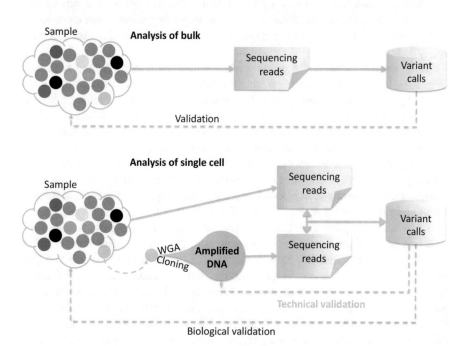

Fig. 3.3 Overview of the strategies for mosaic variants discovery. A sample can be analyzed in bulk (top), but in such a strategy rare variants cannot be discovered. At a standard sequencing coverage of 30X, variants below 20% allele frequency are unlikely to be detectable. However, technical and biological validations are equivalent. In the other strategy, individual nuclei or cells are analyzed (bottom). Here, amplification of a cell's genome by either whole genome amplification (WGA) or cloning is required. Because of this, validation of variant cells in amplified DNA (technical validation) is not equivalent to validation in the original sample (biological validation). However, the latter is challenging for rare variants

main and essential disadvantage of this strategy. Namely, analysis of bulk can discover common variants, but rare variants will likely be missed. The definition of common and rare mosaic variants is similar to that of heritable variants in a population where single nucleotide polymorphisms (SNPs) with population frequency above 1% are considered common (Bodmer and Bonilla 2008). But even for common variants, the sensitivity of their detection is a function of the depth of read coverage – higher sensitivity is enabled by higher coverage. At coverage of 30X, which is widely accepted as adequate to find germline SNPs, only mosaic SNVs with variant allele frequency (VAF) above 5%–10% can be discovered as variants with smaller frequencies are likely to have no supporting reads. Such coverage will also be adequate to find mosaic CNAs and aneuploidies as for them, one can integrate and analyze read coverage across the entire length of the mosaic region. Sequencing at higher coverage will be more sensitive to lower-frequency variants, but it will also be proportionally more expensive. The cost associated with deep sequencing is the other constraint of bulk analysis. Because of that, many studies conducted up to now targeted specific genomic regions and thereby ascertained only a fraction of the human genome, for example, only exons or only L1 retrotransposons. We, however, caution that target approaches have more biases and artifacts in data generation, requiring more careful analysis and thorough validation.

In the second strategy, the genome of an individual nucleus or cell is studied (Fig. 3.3). Its fundamental strength is the suitability to discover variants present in the studied cell, independent of their frequency in a tissue. Namely, the number of reads supporting a mosaic variant in a cell is not related to the variant's frequency of the studied tissue. Theoretically, it is expected to detect all mosaic variants by sequencing a cell to an adequate coverage. Additionally, one can employ to own advantage that mosaic variants in a cell typically reside on only one out of the two alleles present in a cell for most chromosomes (apart from aneuploidies, only sex chromosomes have single haplotypes in males). Consequently, the mosaic variants should be present at ~50% VAF in data from single-cell sequencing experiments. This provides a precise analytical way for distinguishing true variants from false positives of DNA preparation or data generation. Furthermore, this feature makes mosaic variants identical (in terms of frequency characteristics) to the heterozygous germline variants in the same cell, enabling testing, optimizing, refining, validating, and estimating the sensitivity of variant discovery methodology using the germline variants as a reference set. A straightforward study design would be to discover heritable germline variants for a bulk sample, isolate and sequence single cell, calibrate and/or ascertain discovery method based on the heritable variant set for data from each cell, discover variants in each cell, and then conduct validations.

This advantage is also a challenge. Since mosaic and germline variants are indistinguishable in frequency, one must compare a genome of single cells to a genome of some other isogenic tissue or cells in order to discover mosaic variants. But if a variant is present in the compared tissue or cell, it may not be discovered. Apparently, common variants are more likely to be present in compared tissue and multiple cells and therefore may escape detection in such pairwise comparison. Below, we will describe an approach for robust finding of common variants in such a setting. Also note, when discovering variants in bulk, it is not necessary to compare to other tissue since mosaic variants are distinct from germline variants by having a lower frequency. But comparison with other tissue improves the discovery sensitivity toward rare variants, albeit at the same expense of likely missing common variants.

The other fundamental challenge of single-cell studies is that most of the detected variants could be so rare in the primary tissue that they could never be validated (Blokzijl et al. 2016; Bae et al. 2018; Abyzov et al. 2017; Abyzov et al. 2012; Behjati et al. 2014; Lodato et al. 2015). It is also unlikely to detect such variants multiple times in distinct cells. Furthermore, as also discussed below, some techniques for manipulations of DNA from a single cell (e.g., whole-genome amplification [WGA]) require significant enhancements to reduce artifacts into the resulting and sequenced DNA. Such artifacts can mimic a real variant, and a call for a variant may reflect such a confounder rather than a real variant present in the original tissue and analyzed cell/nucleus. These false calls may be still validated in the amplified DNA. Therefore, it is necessary to differentiate between technical validation and validation of variant presence in manipulated DNA (technical validation) and validation of variants in cells from the original tissue sample (biological validation) (Fig. 3.3). Also, such false calls arising from amplification artifacts will be random from cell to cell and will thus look like rare variants. For that reason, in order to have high confidence in and make conclusions about rare variants, one must minimize DNA artifacts created during DNA preparation or at least understand the artifacts and have robust analytics for filtering them out. Finally, analysis of single cells has a fundamental advantage over bulk sequencing for cell lineage tracing – the presence of multiple variants in a cell – and their sharing across multiple cells is easily inferred.

Whole-Genome Sequencing

Whole-genome sequencing (WGS) is the least biased and most comprehensive way to analyze the genome of a bulk or a cell. Analysis of WGS data allows one to study all variant types, albeit with variable efficiency, across the entire genome. So far, studies of mosaic variants with WGS were conducted by using only short read sequencing technologies to keep a lower cost of data generation. The efficiency of detecting mosaic variants in WGS from bulk depends on the coverage depth, while for single cells, it also depends on the uniformity of amplification across the entire genome. As was reasoned above for discovering SNVs (and indels) from bulk, it is necessary to have a coverage of 100X or more. However, 30X–40X coverage of the genome of clones or single cells is sufficient for finding both germline and mosaic variants, as in single cells or clones mosaic variants have characteristics of heterozygous germline variants (Blokzijl et al. 2016; Bae et al. 2018; Abyzov et al. 2017; Lodato et al. 2015; Wang et al. 2012).

The power to detect SVs from WGS depends on physical coverage, i.e., when counting bases in an entire DNA fragment rather than just in sequenced reads (Korbel et al. 2007; Korbel et al. 2009). Therefore, using mate pair libraries with DNA fragments of 2–20 kbp in length allows for an efficient and cost-effective way to detect mosaic SVs as with such libraries, physical coverage is folds larger than sequencing coverage.

Another strength of WGS is that read coverage at a given locus is roughly proportional to its copy number, i.e., a deletion or duplication of a particular region is reflected, respectively, in a decreased or increased coverage, so analysis of the depth of coverage can be used to find CNAs and aneuploidies (Bentley et al. 2008; Abyzov et al. 2011). Although biases in the coverage do exist, they can be corrected for as their sources are known. In single cells with uniform amplification, where the CNA is present at 50% VAF, analysis of coverage can reveal CNAs as small as a few dozen kbp in size even at shallow coverages of 1X–5X (Abyzov et al. 2012; McConnell et al. 2013). Additional separation of reads by DNA strands by means of special library preparation allows for discovering SVs (other than CNA) at extremely shallow coverage (Falconer et al. 2012). However, such library preparation can be conducted for dividing cells only. Regarding discoveries in bulk, a depth-of-coverage approach will only detect frequent CNAs, i.e., present in a large fraction of the cells.

Capture and Sequencing

During the preparation of a sequencing library, one can enrich for particular regions of the genome – target regions – to increase the sequencing coverage of those regions while generating fewer reads, resulting in a cost-effective analytic strategy either in single cells or in bulk. Additionally, higher coverage of targeted regions increases the sensitivity for detecting variants with low frequency in bulk. Enrichment is conducted by hybridizing fragments of DNA from a sample to a set of oligomers synthesized to be complementary to the targeted genomic regions. Specifically, the oligos are complementary to the sequences of the target regions in the reference genome. The oligos with hybridized sample DNA are pulled out of the hybridization reaction using attached baits, typically a biotin moiety. The DNA is then melted, oligomers are removed, and the remaining sample DNA is then prepared as a library and sequenced. It is now a routine practice to use such a capture approach to sequence only the coding portion of the genome – exome sequencing. Exons comprise only about 1% of the entire human genome but harbor most variants with strong functional consequences. The shortcoming of the capture is that it introduces biases into the coverage across the whole genome and between target regions. As a consequence, coverage across the genome is not uniform. While depth-of-coverage analysis is still possible, it is only powerful to find large CNAs in single cells and is likely unsuitable for finding mosaic CNAs in bulk. Additionally, indels within the target regions are poorly captured as the corresponding fragments of DNA are not complementary to the oligomers.

Custom capture libraries can target genes of interest, their exons and promoters, regulatory elements, or any regions of interest including sites where mosaic variants have been discovered. The latter case was proven to be effective for genotyping and confirming across multiple brain regions the presence of mosaic variants discovered from the analysis of clones (Bae et al. 2018). Furthermore, capture libraries can target certain elements in the genome, regardless of their location, as was shown for families of retrotransposons in the human genome (Baillie et al. 2011; Upton et al. 2015). When sequenced, such captured DNA yields reads mapping primarily on or around sequences of retrotransposons across the entire genome, enabling the discovery of variations reflecting insertions of both germline and mosaic retrotransposons. However, it is important to note that when applied to bulk, biases and artifacts during capture can result in false positives that look like low-frequency mosaic insertions. Consequently, the variant cells are to be validated with orthogonal techniques.

Amplicon-Seq and Enrichment for L1 Elements

Amplicon-seq is an approach where one sequences a pool of DNA from multiple (up to a few hundred) PCR reactions, amplifying sequences of distinct target genomic regions; it is an alternative to capture sequencing. Same as capture, it results in an enrichment of targeted regions in the DNA to be sequenced. And since the total length of the regions is manyfold smaller than the length of the entire genome, sequencing only a million reads per amplicon pool already yields an extremely deep coverage of the targets, allowing discovery and confirmation of low-frequency mosaic variants (Martincorena et al. 2015). As individual amplification reaction needs to be conducted for each target region, this approach is used for a relatively few regions at a time, such as the analysis of genomic sequence in coding regions of up to hundreds of genes (Easton et al. 2015). Similar to capture, amplicon-seq can be employed for a genome-wide variant discovery in loci with a particular genomic sequence, like a sequence of retrotransposons. It was shown that semi-targeted PCR of L1 retrotransposon (only one of the primers contains the 3′-end of active L1 elements, while other primers degenerate) allows finding mosaic and previously unknown germline L1 elements in human brain (Evrony et al. 2012; Erwin et al. 2016; Badge et al. 2003). Same as the capture approach, amplicon-seq results in uneven coverage between targeted regions; however, unlike capture, it is well suited to amplify indels.

Its main caveat is that the DNA polymerase introduces errors, such as mismatches and indels, into the amplified DNA, and the errors may look like mosaic variants. To alleviate this drawback, one may use high-fidelity polymerase. In order to confidently detect mosaic variants from the data generated by amplicon-seq, estimation of polymerase's background error rate is necessary (Abyzov et al. 2017).

Another disadvantage of amplicon-seq is that PCR reactions can create chimeric DNA fragments, and chimeras are more likely to form for loci with common repeat sequences in the human genome, including Alu and L1 retroelements. Such chimeras can lead to false calls, particularly for low-frequency mosaic retrotransposon insertions. Furthermore, the lack of a global view of variations in genomes can lead to the misinterpretation of deletions as mosaic L1 insertions (Erwin et al. 2016).

Single-Cell Whole-Genome Amplification

The amount of DNA contained in a cell is roughly 6 ng and is too little to be sequenced by the current sequencing methods and, consequently, needs to be amplified. In vitro whole-genome amplification (WGA) is the most crucial step in the analysis of genomes of a single cell since the quality of the amplified DNA is the major determinant for finding mosaic variants. WGA is conducted using one of several enzymatic protocols that apply DNA polymerases and, recently, RNA polymerases. Two major characteristics that define the quality of WGA are error rate in the amplified DNA and the uniformity of amplification across the genome. While biases in amplification and errors in the amplified DNA are intrinsic to all of existing protocols, they typically vary between different protocols and utilized polymerases. The uniformity of amplification is judged by various metrics, including variance in the distribution of VAFs for heterozygous SNPs (Zhang et al. 2015), the rate of allele dropouts (i.e., fraction of the genome with only one haplotype amplified) (Evrony et al. 2012), median absolute pairwise difference measure (Cai et al. 2014), and variation in the coverage (i.e., variance in read-depth distribution for bins of a particular size across the genome) (Chen et al. 2017). Amplification errors in DNA can be judged from the fraction of read mismatches in sequenced reads (de Bourcy et al. 2014) and the rate of chimeric sequences (Picher et al. 2016).

The earliest method for WGA is degenerate oligonucleotide-*primed* PCR (DOP-PCR) (Telenius et al. 1992). Still now, this popular method and its modification result in the most uniform coverage across the genome at a larger scale (i.e., few-dozen kbp). Because of that, it is best suited for finding CNAs and chromosomal aneuploidies (McConnell et al. 2013; Cai et al. 2014; Knouse et al. 2014; Navin et al. 2011; Gawad et al. 2016). Another widely used method is multiple displacement amplification (MDA), which conducts amplification at a constant temperature by using the *ϕ29* polymerase (Dean et al. 2002). Owing to high processivity and fidelity of the polymerase, this method has the advantage over DOP-PCR by producing much longer – up to several kbp – DNA fragments and having a much lower base substitution rate in the produced DNA. Because of these properties, DNA from MDA is well suited for the discovery of SNVs, indels, and MEIs (Lodato et al. 2018; Evrony et al. 2012; Lodato et al. 2015). On the other hand, due to its suffering from a relatively high rate of allelic dropouts and generally nonuniform coverage, discovering CNAs and aneuploidies from MDA amplified cells is problematic (Cai et al. 2014). Still, since theoretically MDA can enable ascertainment of all variant types in a cell, it is viewed as the most promising amplification method. Its main weakness is the exponential nature of amplification resulting in an uneven coverage and leading to errors in the first steps of amplification being propagated at high frequency into the resulting DNA. Similarly, the amplification propagates unrepaired DNA damage in the original DNA as error into the amplified DNA. Ongoing developments to improve the method are mostly focused on mitigating these weaknesses.

A quasilinear amplification can be conducted with multiple annealing and looping-based amplification cycles (MALBAC) (Zong et al. 2012). Following the protocol of the method, one is able to conduct up to five cycles of linear amplification from the original cell's DNA by utilizing the combination of special primers and optimized temperatures for DNA melting, looping, and amplification. After this pre-amplification, one conducts an exponential amplification phase the same as in regular MDA. It was found that the addition of the single-stranded DNA binding protein from *Thermus thermophilus* HB8 improved the efficiency of MDA amplification (Inoue et al. 2006). Recently, it was proposed to improve MDA by utilizing a DNA primase from *Thermus thermophilus* that would prime amplification strand instead of random primers (Picher et al. 2016). The innovation was reported to result in fewer allelic dropouts, a more uniform coverage, and improved SNV detections (Picher et al. 2016). Improvement in the amplification uniformity was observed from limiting reaction volume (Hutchison et al. 2005; Marcy et al. 2007; Gole et al. 2013; Fu et al. 2015) and from the addition of proper concentration of trehalose (Pan et al. 2008). Recently, described Linear Amplification via Transposon Insertion (LIANTI) conducts linear amplification by means of T7 RNA polymerase (Chen et al. 2017). As apparent from the name of the utilized enzyme, the amplified material is RNA, which is, at the final step of the protocol, being reverse-transcribed into the DNA.

The same work has also proposed a solution to the dominant amplification error of cytosine to tyrosine. The prominent source of the errors is cytosine deamination into uracil upon cell lysis, which is read as tyrosine when copied by a polymerase or being sequenced. To a lesser extent, deamination is caused by a natural process that occurs at a low rate randomly in the genome (Shen et al. 1994). The proposed solution is based on treating DNA from lysed cells with uracil-DNA glycosylase, which eliminated uracil-deaminated bases (Chen et al. 2017). In parallel, the errors were shown to be diminished or entirely eliminated by performing cell lysis and DNA denaturation on ice through alkaline lysis prior to conducting MDA (Dong et al. 2017). In particular, when using human primary fibroblasts, the frequency of substitutions of called SNVs in amplified single cells closely resembled those in called SNVs in clonally expanded colonies (Dong et al. 2017).

All these improvements pave the way for precise and complete variant calling in a single cell, potentially enabling such discoveries now. However, we caution the reader that many of the recent modifications to amplification protocols have not yet been independently verified. Additionally, it is not clear how and whether the amplification outcomes of various protocols depend on cell type, nuclei, and different prior conditions of sample collection, storage, and cell/nuclei extraction.

Single-Cell Clonal Expansion

Currently, single-cell WGA is far less faithful than genome duplication in a dividing cell. Instead of just polymerase, cells employ a sophisticated molecular machinery to minimize error during copying their genome, as well as to proofread and correct error. The strategy of single-cell clonal expansion leverages the highly precise DNA amplification in proliferating cells. Specifically, cells extracted from a sample are cultured until the size of such clonal colony (i.e., the number of cells) is large enough to extract the necessary amount of DNA for sequencing (Blokzijl et al. 2016; Bae et al. 2018; Abyzov et al. 2017; Saini et al. 2016; Abyzov et al. 2012; Behjati et al. 2014) (Fig. 3.3). Cells of each colony will have the genome of the founder cell cloned, thereby bypassing challenges of WGA. Each cell may also have variants created during culture, which can be distinguished from the genuine mosaic variant of the founder cell by their frequency. Ideally, if all cells in a colony proliferate at the same rate and do not die, then variants created in culture will have a small VAF in the colony except for those created during the first few divisions. For the division of the founder cell (i.e., first division in colony), created variants will be present on only one haplotype out of four for diploid chromosomes and out of two for haploid chromosomes after genome duplication. Therefore, when haplotypes segregate into daughter cells, the allele frequency of created variants in the colony will be 25% and 50%, respectively. In the ideal case, these values will be propagated through all later division and will be significantly lower than 50% and 100% allele frequency for genuine mosaic variants on, respectively, diploid and haploid chromosomes when DNA is harvested. In reality, however, cells divide at different rates, die, and senescence, increasing frequency of some culture-introduced variants and making them less distinct from mosaic variants in the founder cell. Higher sequencing coverage would allow for a finer distinction between small deviations from 50% (or 100%) VAF, but that would also make experiments more expensive. Therefore, monitoring early stages of clonal expansion and checking that there was no disparity at the start of cell proliferation are likely to ensure clonality of the produced colonies – the key assumption in the discussed approach. In many current studies, clonality is presumed as very likely (Abyzov et al. 2012) or verified from the obtained sequencing data (Blokzijl et al. 2016; Bae et al. 2018).

The fundamental limitation of the clonal expansion is that it relies on proliferation potential of the cell. This makes it hardly applicable to study terminally differentiated cells, like neurons, and undifferentiated cells with mosaic mutations, preventing their proliferation. A notable exception is cells that can be reprogrammed into induced pluripotent state and then cultured (Abyzov et al. 2012). Currently, several cell types such as fibroblasts, keratinocytes, blood cells, and renal epithelial cells can be reprogrammed into induced pluripotent stem cells (iPSC) (Zhou et al. 2011; Zhou et al. 2012). However, a possibility to make a colony out of a cell may also depend on the cell's mosaic mutations. Consequently, the mutations making a cell unculturable or not amendable to reprogramming cannot be detected, and this results in an intrinsic bias of the strategy for discovering mosaic variants from clonal expansion.

Other Strategies

To mitigate and overcome limitations of WGA and clonal expansion, a few strategies have been developed and described. Perhaps the most prominent one, particularly for the analysis of mosaicism in the brain, is an extension of clonal expansion for terminally differentiated somatic cells – Somatic Cell Nuclear Transfer (SCNT) (Hazen et al. 2016). In this strategy, to clonally expand a somatic cell of a mouse, the cell's nucleus is transferred to an enucleated oocyte. When implanted into the uterus, the oocyte can develop into a living mouse pup. DNA collected from any tissue of the animal will represent the genome of the originally transferred nucleus with the addition of mosaic variants occurring in the development of the animal, referred to above as created during culture. As reasoned in the previous section, it is straightforward to distinguish such variants by their allele frequency. As this strategy was successfully applied to study mosaicism in mouse brain, it is likely to be applicable to other mammals and animals. However, using it in humans is unethical. Furthermore, this strategy still is not able to discover mutations that make cells unculturable, like mutations that prevent a cell from replication.

Another hybrid strategy mitigates the challenges of WGA for cells with limited proliferation potential. It starts from a single cell, arrests its division at S-phase, and proceeds with WGA from the still-undivided cell (Leung et al. 2015). The advantage is that WGA starts from a larger DNA amount, improving the quality and uniformity of amplified DNA. The strategy could be beneficial for studying mosaicism in cells that may not proliferate long enough to be extended into a colony, such as glial cells.

Analytics of Variant Discovery

Concept of Variant Discovery

Once sequencing data has been generated, it has to be analyzed to discover variants. The most common workflow for this analysis consists of the two major steps: (i) aligning of sequencing reads to a reference genome, called mapping, and (ii) finding systematic abnormalities in the data to generate variant calls. Mappings are done with dedicated software that assumes that sequenced reads map somewhere in the reference with a high 95%–99% sequence identity. Such a valid assumption, as genomic diversity in the human population is less than 0.5% (Chaisson et al. 2019), allows for fast read mapping. Still, billions of WGS reads per personal genome take hours to map even when utilizing dozens of CPUs. Systematic abnormalities could be the same mismatch at the same position across multiple reads – suggesting SNV, abnormally large distances between pairs of read at a certain genomic location (fragment size is defined by prepared sequencing library) – suggesting a deletion and others (McKenna et al. 2010; Medvedev et al. 2009). As a result, personal variants are called relative to the reference genome. Therefore, variants that have the strongest signal in the data (such as germline variants, particularly those that are homozygous) are easier to discover. Variants existing in a subset of sequenced alleles, such as mosaic variants in a tissue, have weaker signals and are hard to distinguish from false positives. To boost the power in discovering low-frequency variants, it is common to perform variant discovery in the sample of interest relative to some other reference sample from the same person. Such a comparison is widely used in the field of cancer genomics where a sample of cancer is compared against some normal sample of the same individual. Still, the comparison is made through sequencing of each sample and mapping reads to the reference genome.

An alternative to reference-based read alignment is assembly-based variant discovery where reads are first assembled into long contigs and then compared to the reference genome. While this approach is theoretically more accurate, it is also computationally much more intensive. Therefore, the standard in the field is to first map reads to the reference and then perform local assembly around the sites of suspected variants.

Leveraging Analytics from Cancer Genomics

As discussed, finding mosaic variants is in many ways similar to finding somatic variants in cancers, and therefore, the wealth of analytics developed in cancer genomics for pairwise sample comparison can be applied (Abyzov et al. 2012; Koboldt et al. 2012; Cibulskis et al. 2013; Wala et al. 2018; Kim et al. 2018). However, there are essential differences. When discovering variants from single-cell analysis, variant calls with measured low frequency in data from each cell will likely represent amplification, or culturing artifacts, are not of interest and should be filtered out. Next, pairwise comparison of bulk samples will miss mosaic variants with high VAF in the reference tissue as such variants are hard to distinguish from germline variants. In cancer genomics, according to the main paradigm of clonal cancer evolutions, driver mutations are absent in normal tissues and only observable in cancer samples. In contrast, mosaic variants with high frequency in a tissue are likely to exhibit the largest phenotypic effect on the tissue (or organism) and thus could be of the highest importance to detect. Besides, higher-frequency variants are the most informative markers for lineage tracing (Bae et al. 2018; Evrony et al. 2015; Behjati et al. 2014; Lodato et al. 2015; Lee-Six et al. 2018), which can inform about development (Bae et al. 2018) and clonality in a tissue (Abyzov et al. 2017; Lodato et al. 2015; Lee-Six et al. 2018). Therefore, paying attention to high-frequency variants is of paramount importance in discovering mosaic variants. However, what to consider as high frequency can be different in each study case.

Detecting High-Frequency Variants

A few high-frequency variants can be detected when comparing genome of a cell to a tissue. Specifically, their discovery is possible if the variants have been sampled in multiple cells and gained no or little support in the read data for tissue (Bae et al. 2018; Abyzov et al. 2012; Lodato et al. 2015). Because of the latter, such variants would typically have 1% allele frequency or lower in tissues, i.e., be on the lower end of the frequency spectrum for high-frequency variants. Increasing sequencing depth would allow for a finer distinction of germline and high-frequency mosaic variants at the expense of higher experimental cost. For instance, at WGS of 80X for both single cells and the compared tissue, a mosaic SNV with a VAF of 20% is distinguished from 50% VAF of germline SNP with a p-value of about 10^{-4}. But since the individual genome has roughly two million heterozygous SNPs (1000 Genomes Project Consortium et al. 2015), such statistical confidence is not significant when adjusting for multiple hypothesis (i.e., the number of the SNPs) testing.

Three strategies to solve this issue have been proposed. In the first and, likely, the most comprehensive strategy, instead of comparing to tissues from the same person, a cell's genome is compared to genomes of a twin or the parents of the individual from whom the cell was obtained (Abyzov et al. 2017). Such a comparison would eliminate all inherited germline variants, calling mosaic and germline de novo variants in each cell. Germline de novo variants can then be identified as being present in all cells. The obvious disadvantage of this strategy is that parental genomes are rarely available along with the primer tissue for the analysis. In the second strategy, the genome of each cell is compared to some other unrelated genome (Lee-Six et al. 2018). In this case, germline variants different between the two genomes are called alone with mosaic variants. While filtering out germline variants is also straightforward – they must be present in (almost) all cells – the large count of different germline variants in unrelated individuals (over a million) makes the filtering prone to errors. Besides, comparing unrelated genomes is likely to generate more false calls due to violation of the basic assumption that compared genomes are mostly the same (Fig. 3.4). The last strategy mitigates disadvantages for the first two by conducting a complete comparison between all sequenced cells from an individual, so called cell-to-cell comparison (Bae et al. 2018). In this strategy, the same variant can be called multiple times from different compared pairs of cells. By analyzing how consistently the same variant is called for the same cell or a group of cells, one can distinguish mosaic variants, germline SNPs, and false positives (Bae et al. 2018). The major limitation of such comparison is that it requires a number of comparisons that is roughly squared to the number of available cells. For a larger number of cells sequenced, such computations may become

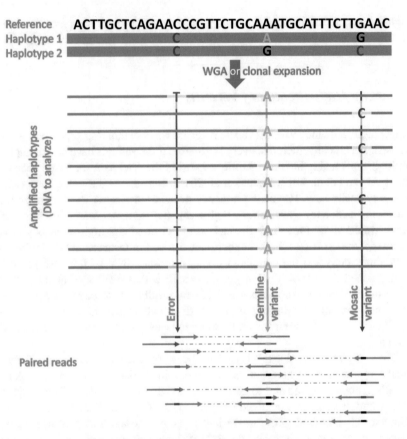

Fig. 3.4 Conceptual depiction of resolving true mosaic SNV from WGA error by haplotype phasing. A cell or a nucleus has heterozygous germline SNP A > G, a nearby true mosaic SNVs G>C (in blue), and a nearby amplification error C>T (in red). Two haplotypes in a cell/nucleus are unevenly amplified so there is a larger count of haplotypes with A-allele of the SNP in the analyzed DNA. Regardless of bias in amplification, the C-allele of true mosaic SNVs will always be in phase with one (G-allele) and out of phase with the other on (A-allele) alleles of the SNP, i.e., there will always be two haplotypes. An amplification error, on the other hand, will sometimes co-occur with one allele (A-allele) and be out of phase with the other one (G-allele) of the SNP, i.e., there will be three haplotypes existing in DNA. Such haplotype structure will be propagated into sequenced reads and can be reconstructed from it. Note that in this example, the measured VAFs for the mosaic SNVs are lower than that of the error, yet haplotype reconstruction will differentiate between the two. Also note that a WGA error can also be present on two haplotypes (not shown)

prohibitively long. So far, the advantage of complete cell-to-cell comparison was demonstrated for finding SNVs from sequencing data from clones, but conceptually, it should be applicable to finding all variants both from clones and from single cells. But we also note that the approach is not capable or intended to distinguish amplification errors from genuine mosaic variants.

Resolving WGA Errors

Filtering out false positives due to base substitution errors during WGA can be partially accomplished by setting a cutoff on VAF in the sequencing data from a cell. For variants discovered from sequencing data for clones, the measured VAF is tight already at 30X coverage, with standard deviation reflecting stochastic coverage fluctuations. In such a case, the VAF cutoff can efficiently separate mosaic SNVs and germline heterozygous SNPs from culturing artifacts. However, in WGA data, measured VAF is over-dispersed primarily due to uneven amplification. In such a case, distributions of VAF for culturing artifacts and for mosaic SNVs and germline heterozygous SNPs overlap significantly, and applying VAF cutoff is not that effective. An alternative approach is to determine the haplotype of the candidate mosaic variants. A real mosaic variant must be on one and only one haplotype and, by extrapolation, on all copies of the haplotype generated by WGS. When analyzing corresponding sequencing data, such a variant must be perfectly *in phase* with nearby heterozygous germline allele, if such allele exists (Fig. 3.4). In other words, the data should suggest the existence of only two haplotypes: one with a variant and another one without it. Note, only heterozygous germline variants allow for distinguishing two haplotypes in a cell. Errors of WGA can be on both or only one haplotype; however, on one haplotype, they (the errors) are likely to be only on a fraction of the haplotype's copies, as they occurred during amplification, rather than being constitutive to the haplotype before amplification. These conditions will result in poor phasing for WGA errors. A remarkable advantage over imposing VAF cutoff is that this approach enables confident distinction between real mosaic variants and an amplification error even if the frequency of the former is lower than the frequency of the latter (Fig. 3.4).

Theoretically, the approach is applicable to all variant types including indels, SVs, CNAs, and MEIs, but its utility so far has been demonstrated for mosaic SNVs (Lodato et al. 2018; Ju et al. 2017; Freed and Pevsner 2016). Furthermore, the efficiency of the approach depends on the number of heterozygous SNPs in a personal genome and on the length of the sequencing reads/fragments. There are approximately four million germline variants per human genome, with less than three million being heterozygous (1000 Genomes Project Consortium et al. 2015). Consequently, on average, each human has a heterozygous variant per 500–2000 bases. Existing short read next-generation technologies sequence fragments of about 450 bp in length, allowing for confident haplotype resolution for a limited fraction of only about ~20% of candidate mosaic variants (Lodato et al. 2018). Therefore, read/fragment length represents the major challenge of the haplotype-phasing approach for filtering out false positives. It can be anticipated that long reads generated by Oxford Nanopore Technologies and Pacific Biosciences will allow for utilizing the approach to its full potential; however, as discussed above, long reads are two to ten times more expensive per sequenced base and suffer from excessive sequencing error rate. A possible and somewhat intermediate solution could be sequencing with linked short reads by 10X genomics that can be used to phase variants into long haplotypes of millions of bases in length (Chaisson et al. 2019; Kitzman 2016), thereby promising near complete phasing of mosaic variants. However, the applicability and efficiency of this sequencing technology for finding mosaic variants in with WGA DNA has not yet been demonstrated.

References

1000 Genomes Project Consortium (2015) A global reference for human genetic variation. Nature. Nature Publishing Group 526(7571):68–74. PMCID: PMC4750478

Abyzov A, Urban AE, Snyder M, Gerstein M (2011) CNVnator: an approach to discover, genotype, and characterize typical and atypical CNVs from family and population genome sequencing. Genome Res. Cold Spring Harbor Lab 21(6):974–984. PMCID: PMC3106330

Abyzov A, Mariani J, Palejev D, Zhang Y, Haney MS, Tomasini L, Ferrandino AF, Rosenberg Belmaker LA, Szekely A, Wilson M, Kocabas A, Calixto NE, Grigorenko EL, Huttner A, Chawarska K, Weissman S, Urban AE, Gerstein M, Vaccarino FM (2012) Somatic copy number mosaicism in human skin revealed by induced pluripotent stem cells. Nature 492(7429):438–442. PMCID: PMC3532053

Abyzov A, Tomasini L, Zhou B, Vasmatzis N, Coppola G, Amenduni M, Pattni R, Wilson M, Gerstein M, Weissman S, Urban AE, Vaccarino FM (2017) One thousand somatic SNVs per skin fibroblast cell set baseline of mosaic mutational load with patterns that suggest proliferative origin. Genome Res 27(4):512–523. PMCID: PMC5378170

Alexandrov LB, Jones PH, Wedge DC, Sale JE, Campbell PJ, Nik-Zainal S, Stratton MR (2015) Clock-like mutational processes in human somatic cells. Nat Genet. Nature Publishing Group 47(12):1402–1407. PMCID: PMC4783858

Alexandrov LB, Nik-Zainal S, Wedge DC, Aparicio SAJR, Behjati S, Biankin AV, Bignell GR, Bolli N, Borg A, Børresen-Dale A-L, Boyault S, Burkhardt B, Butler AP, Caldas C, Davies HR, Desmedt C, Eils R, Eyfjörd JE, Foekens JA, Greaves M, Hosoda F, Hutter B, Ilicic T, Imbeaud S, Imielinski M, Imielinsk M, Jäger N, Jones DTW, Jones D, Knappskog S, Kool M, Lakhani SR, López-Otín C, Martin S, Munshi NC, Nakamura H, Northcott PA, Pajic M, Papaemmanuil E, Paradiso A, Pearson JV, Puente XS, Raine K, Ramakrishna M, Richardson AL, Richter J, Rosenstiel P, Schlesner M, Schumacher TN, Span PN et al (2013) Signatures of mutational processes in human cancer. Nature 500:415–421. PMCID: PMC3776390

Aubert G, Lansdorp PM (2008) Telomeres and aging. Physiol Rev Am Physioll Soc 88(2):557–579

Bacolla A, Cooper DN, Vasquez KM (2014) Mechanisms of base substitution mutagenesis in cancer genomes. Genes (Basel). Multidisciplinary Digital Publishing Institute 5(1):108–146. PMCID: PMC3978516

Badge RM, Alisch RS, Moran JV (2003) ATLAS: a system to selectively identify human-specific L1 insertions. Am J Hum Genet 72(4):823–838. PMCID: PMC1180347

Bae T, Tomasini L, Mariani J, Zhou B, Roychowdhury T, Franjic D, Pletikos M, Pattni R, Chen B-J, Venturini E, Riley-Gillis B, Sestan N, Urban AE, Abyzov A, Vaccarino FM (2018) Different mutational rates and mechanisms in human cells at pregastrulation and neurogenesis. Science. American Association for the Advancement of Science 359(6375):550–555. PMCID: PMC6311130

Baillie JK, Barnett MW, Upton KR, Gerhardt DJ, Richmond TA, De Sapio F, Brennan PM, Rizzu P, Smith S, Fell M, Talbot RT, Gustincich S, Freeman TC, Mattick JS, Hume DA, Heutink P, Carninci P, Jeddeloh JA, Faulkner GJ (2011) Somatic retrotransposition alters the genetic landscape of the human brain. Nature 479(7374):534–537. PMCID: PMC3224101

Behjati S, Huch M, van Boxtel R, Karthaus W, Wedge DC, Tamuri AU, Martincorena I, Petljak M, Alexandrov LB, Gundem G, Tarpey PS, Roerink S, Blokker J, Maddison M, Mudie L, Robinson B, Nik-Zainal S, Campbell P, Goldman N, van de Wetering M, Cuppen E, Clevers H,

Stratton MR (2014) Genome sequencing of normal cells reveals developmental lineages and mutational processes. Nature 513(7518):422–425. PMCID: PMC4227286

Bentley DR, Balasubramanian S, Swerdlow HP, Smith GP, Milton J, Brown CG, Hall KP, Evers DJ, Barnes CL, Bignell HR, Boutell JM, Bryant J, Carter RJ, Keira Cheetham R, Cox AJ, Ellis DJ, Flatbush MR, Gormley NA, Humphray SJ, Irving LJ, Karbelashvili MS, Kirk SM, Li H, Liu X, Maisinger KS, Murray LJ, Obradovic B, Ost T, Parkinson ML, Pratt MR, IMJ R, Reed MT, Rigatti R, Rodighiero C, Ross MT, Sabot A, Sankar SV, Scally A, Schroth GP, Smith ME, Smith VP, Spiridou A, Torrance PE, Tzonev SS, Vermaas EH, Walter K, Wu X, Zhang L, Alam MD, Anastasi C et al (2008) Accurate whole human genome sequencing using reversible terminator chemistry. Nature 456(7218):53–59

Blokzijl F, de Ligt J, Jager M, Sasselli V, Roerink S, Sasaki N, Huch M, Boymans S, Kuijk E, Prins P, Nijman IJ, Martincorena I, Mokry M, Wiegerinck CL, Middendorp S, Sato T, Schwank G, Nieuwenhuis EES, Verstegen MMA, van der Laan LJW, de Jonge J, IJzermans JNM, Vries RG, van de Wetering M, Stratton MR, Clevers H, Cuppen E, van Boxtel R (2016) Tissue-specific mutation accumulation in human adult stem cells during life. Nature 538(7624):260–264

Bodmer W, Bonilla C (2008) Common and rare variants in multifactorial susceptibility to common diseases. Nat Genet. Nature Publishing Group 40(6):695–701. PMCID: PMC2527050

Bruce AW, Zernicka-Goetz M (2010) Developmental control of the early mammalian embryo: competition among heterogeneous cells that biases cell fate. Curr Opin Genet Dev 20(5):485–491

C. elegans Sequencing Consortium (1998) Genome sequence of the nematode C. elegans: a platform for investigating biology. Science. American Association for the Advancement of Science 282(5396):2012–2018

Cai X, Evrony GD, Lehmann HS, Elhosary PC, Mehta BK, Poduri A, Walsh CA (2014) Single-cell, genome-wide sequencing identifies clonal somatic copy-number variation in the human brain. Cell Rep 8(5):1280–1289. PMCID: PMC4272008

Catcheside DG, Lea DE, Genetics DLJO (1945) Dominant lethals and chromosome breaks in ring-X chromosomes of Drosophila melanogaster. J Genet

Chaisson MJP, Sanders AD, Zhao X, Malhotra A, Porubský D, Rausch T, Gardner EJ, Rodriguez OL, Guo L, Collins RL, Fan X, Wen J, Handsaker RE, Fairley S, Kronenberg ZN, Kong X, Hormozdiari F, Lee D, Wenger AM, Hastie AR, Antaki D, Anantharaman T, Audano PA, Brand H, Cantsilieris S, Cao H, Cerveira E, Chen C, Chen X, Chin C-S, Chong Z, Chuang NT, Lambert CC, Church DM, Clarke L, Farrell A, Flores J, Galeev T, Gorkin DU, Gujral M, Guryev V, Heaton WH, Korlach J, Kumar S, Kwon JY, Lam ET, Lee JE, Lee J, Lee W-P, Lee SP et al (2019) Multi-platform discovery of haplotype-resolved structural variation in human genomes. Nat Commun. Nature Publishing Group 10(1):1784. PMCID: PMC6467913

Chen C, Xing D, Tan L, Li H, Zhou G, Huang L, Xie XS (2017) Single-cell whole-genome analyses by Linear Amplification via Transposon Insertion (LIANTI). Science. American Association for the Advancement of Science 356(6334):189–194

Chronister WD, Burbulis IE, Wierman MB, Wolpert MJ, Haakenson MF, Smith ACB, Kleinman JE, Hyde TM, Weinberger DR, Bekiranov S, McConnell MJ (2019) Neurons with complex karyotypes are rare in aged human Neocortex. Cell Rep 26(4):825–827

Cibulskis K, Lawrence MS, Carter SL, Sivachenko A, Jaffe D, Sougnez C, Gabriel S, Meyerson M, Lander ES, Getz G (2013) Sensitive detection of somatic point mutations in impure and heterogeneous cancer samples. Nat Biotechnol 31(3):213–219. PMCID: PMC3833702

Conklin EG (1905) The organization and cell-lineage of the ascidian egg. 13

Conrad DF, Bird C, Blackburne B, Lindsay S, Mamanova L, Lee C, Turner DJ, Hurles ME (2010) Mutation spectrum revealed by breakpoint sequencing of human germline CNVs. Nat Genet 42(5):385–391

De S (2011) Somatic mosaicism in healthy human tissues. Trends Genet 27(6):217–223

de Bourcy CFA, De Vlaminck I, Kanbar JN, Wang J, Gawad C, Quake SR (2014) A quantitative comparison of single-cell whole genome amplification methods. Wang K, editor. PLoS ONE 9(8):e105585. PMCID: PMC4138190

Dean FB, Hosono S, Fang L, Wu X, Faruqi AF, Bray-Ward P, Sun Z, Zong Q, Du Y, Du J, Driscoll M, Song W, Kingsmore SF, Egholm M, Lasken RS (2002) Comprehensive human genome amplification using multiple displacement amplification. Proc Natl Acad Sci USA 99(8):5261–5266. PMCID: PMC122757

Dong X, Zhang L, Milholland B, Lee M, Maslov AY, Wang T, Vijg J (2017) Accurate identification of single-nucleotide variants in whole-genome-amplified single cells. Nat Methods. Nature Research. PMCID: PMC5408311

Duncan AW, Hanlon Newell AE, Smith L, Wilson EM, Olson SB, Thayer MJ, Strom SC, Grompe M (2012) Frequent aneuploidy among normal human hepatocytes. Gastroenterology 142(1):25–28. PMCID: PMC3244538

Easton DF, Pharoah PDP, Antoniou AC, Tischkowitz M, Tavtigian SV, Nathanson KL, Devilee P, Meindl A, Couch FJ, Southey M, Goldgar DE, Evans DGR, Chenevix-Trench G, Rahman N, Robson M, Domchek SM, Foulkes WD (2015) Gene-panel sequencing and the prediction of breast-cancer risk. N Engl J Med 372(23):2243–2257. PMCID: PMC4610139

Erwin JA, Paquola ACM, Singer T, Gallina I, Novotny M, Quayle C, Bedrosian TA, Alves FIA, Butcher CR, Herdy JR, Sarkar A, Lasken RS, Muotri AR, Gage FH (2016) L1-associated genomic regions are deleted in somatic cells of the healthy human brain. Nat Neurosci. Nature Publishing Group 19(12):1583–1591. PMCID: PMC5127747

Evrony GD, Cai X, Lee E, Hills LB, Elhosary PC, Lehmann HS, Parker JJ, Atabay KD, Gilmore EC, Poduri A, Park PJ, Walsh CA (2012) Single-neuron sequencing analysis of l1 retrotransposition and somatic mutation in the human brain. Cell 151(3):483–496. PMCID: PMC3567441

Evrony GD, Lee E, Mehta BK, Benjamini Y, Johnson RM, Cai X, Yang L, Haseley P, Lehmann HS, Park PJ, Walsh CA (2015) Cell lineage analysis in human brain using endogenous retroelements. Neuron 85(1):49–59. PMCID: PMC4299461

Falconer E, Hills M, Naumann U, Poon SSS, Chavez EA, Sanders AD, Zhao Y, Hirst M, Lansdorp PM (2012) DNA template strand sequencing of single-cells maps genomic rearrangements at high resolution. Nat Methods 9(11):1107–1112

Feil R, Wagner J, Metzger D, Chambon P (1997) Regulation of Cre recombinase activity by mutated estrogen receptor ligand-binding domains. Biochem Biophys Res Commun 237(3):752–757

Ferrús A, Garcia-Bellido A (1977) Minute mosaics caused by early chromosome loss. Wilehm Roux Arch Dev Biol Springer-Verlag 183(4):337–349

Forsberg LA, Rasi C, Razzaghian HR, Pakalapati G, Waite L, Thilbeault KS, Ronowicz A, Wineinger NE, Tiwari HK, Boomsma D, Westerman MP, Harris JR, Lyle R, Essand M, Eriksson F, Assimes TL, Iribarren C, Strachan E, O'Hanlon TP, Rider LG, Miller FW, Giedraitis V, Lannfelt L, Ingelsson M, Piotrowski A, Pedersen NL, Absher D, Dumanski JP (2012) Age-related somatic structural changes in the nuclear genome of human blood cells. Am J Hum Genet 90(2):217–228. PMCID: PMC3276669

Franco I, Johansson A, Olsson K, Vrtačnik P, Lundin P, Helgadottir HT, Larsson M, Revêchon G, Bosia C, Pagnani A, Provero P, Gustafsson T, Fischer H, Eriksson M (2018) Somatic mutagenesis in satellite cells associates with human skeletal muscle aging. Nat Commun. Nature Publishing Group 9(1):800. PMCID: PMC5824957

Frank E, Sanes JR (1991) Lineage of neurons and glia in chick dorsal root ganglia: analysis in vivo with a recombinant retrovirus. Development 111(4):895–908

Freed D, Pevsner J (2016) The Contribution of Mosaic Variants to Autism Spectrum Disorder. Bucan M, editor. PLoS Genet. Public Library of Science 12(9):e1006245. PMCID: PMC5024993

Friedberg EC (2003) DNA damage and repair. Nature. Nature Publishing Group 421(6921):436–440

Fu Y, Li C, Lu S, Zhou W, Tang F, Xie XS, Huang Y (2015) Uniform and accurate single-cell sequencing based on emulsion whole-genome amplification. Proc Natl Acad Sci USA. National Acad Sciences 112(38):11923–11928. PMCID: PMC4586872

Garcia-Bellido A. Cell Lineages and Genes (1985) Philosophical transactions of the Royal Society B: Biological Sciences. The Royal Society London 312(1153):101–128

Garcia-Bellido A, Merriam JR (1969) Cell lineage of the imaginal discs in Drosophila gynandromorphs. J. Exp. Zool. John Wiley & Sons. Ltd 170(1):61–75

Garcia-Bellido A, Ripoll P, Morata G (1973) Developmental compartmentalisation of the wing disk of Drosophila. Nature New Biol Nature Publishing Group 245(147):251–253

Gawad C, Koh W, Quake SR (2016) Single-cell genome sequencing: current state of the science. Nat Rev Genet Nature Research 17(3):175–188

Gole J, Gore A, Richards A, Chiu Y-J, Fung H-L, Bushman D, Chiang H-I, Chun J, Lo Y-H, Zhang K (2013) Massively parallel polymerase cloning and genome sequencing of single cells using nanoliter microwells. Nat Biotechnol. Nature Research 31(12):1126–1132. PMCID: PMC3875318

Gossen M, Bujard H (1992) Tight control of gene expression in mammalian cells by tetracycline-responsive promoters. Proc Natl Acad Sci USA. National Academy of Sciences 89(12):5547–5551. PMCID: PMC49329

Harrison DA, Perrimon N (1993) Simple and efficient generation of marked clones in Drosophila. Curr Biol 3(7):424–433

Hazen JL, Faust GG, Rodriguez AR, Ferguson WC, Shumilina S, Clark RA, Boland MJ, Martin G, Chubukov P, Tsunemoto RK, Torkamani A, Kupriyanov S, Hall IM, Baldwin KK (2016) The complete genome sequences, unique mutational spectra, and developmental potency of adult neurons revealed by cloning. Neuron 89(6):1223–1236. PMCID: PMC4795965

Hotta Y, Benzer S (1973) Mapping of behavior in Drosophila mosaics. Symp Soc Dev Biol 31:129–167

Huang AY, Yang X, Wang S, Zheng X, Wu Q, Ye AY, Wei L (2018) Distinctive types of postzygotic single-nucleotide mosaicisms in healthy individuals revealed by genome-wide profiling of multiple organs. Greally JM, editor. PLoS Genet 14(5):e1007395. PMCID: PMC5969758

Hutchison CA, Smith HO, Pfannkoch C, Venter JC (2005) Cell-free cloning using φ29 DNA polymerase. Proc Natl Acad Sci USA. National Academy of Sciences 102(48):17332–17336

Ikehata H, Ono T (2011) The mechanisms of UV mutagenesis. J Radiat Res 52(2):115–125

Inoue J, Shigemori Y, Mikawa T (2006) Improvements of rolling circle amplification (RCA) efficiency and accuracy using Thermus thermophilus SSB mutant protein. Nucleic Acids Res 34(9):e69–e69

Jacobs KB, Yeager M, Zhou W, Wacholder S, Wang Z, Rodríguez-Santiago B, Hutchinson A, Deng X, Liu C, Horner M-J, Cullen M, Epstein CG, Burdett L, Dean MC, Chatterjee N, Sampson J, Chung CC, Kovaks J, Gapstur SM, Stevens VL, Teras LT, Gaudet MM, Albanes D, Weinstein SJ, Virtamo J, Taylor PR, Freedman ND, Abnet CC, Goldstein AM, Hu N, Yu K, Yuan J-M, Liao L, Ding T, Qiao Y-L, Gao Y-T, Koh W-P, Xiang Y-B, Tang Z-Z, Fan J-H, Aldrich MC, Amos C, Blot WJ, Bock CH, Gillanders EM, Harris CC, Haiman CA, Henderson BE, Kolonel LN, Le Marchand L et al (2012) Detectable clonal mosaicism and its relationship to aging and cancer. Nat Genet 44(6):651–658. PMCID: PMC3372921

Jones KT (2008) Meiosis in oocytes: predisposition to aneuploidy and its increased incidence with age. Hum Reprod Update 14(2):143–158

Ju YS, Martincorena I, Gerstung M, Petljak M, Alexandrov LB, Rahbari R, Wedge DC, Davies HR, Ramakrishna M, Fullam A, Martin S, Alder C, Patel N, Gamble S, O'Meara S, Giri DD, Sauer T, Pinder SE, Purdie CA, Borg A, Stunnenberg H, van de Vijver M, Tan BKT, Caldas C, Tutt A, Ueno NT, van Veer LJ, JWM M, Sotiriou C, Knappskog S, Span PN, Lakhani SR, Eyfjörd JE, Børresen-Dale A-L, Richardson A, Thompson AM, Viari A, Hurles ME, Nik-Zainal S, Campbell PJ, Stratton MR (2017) Somatic mutations reveal asymmetric cellular dynamics in the early human embryo. Nature 6:216

Kim S, Scheffler K, Halpern AL, Bekritsky MA, Noh E, Källberg M, Chen X, Kim Y, Beyter D, Krusche P, Saunders CT (2018) Strelka2: fast and accurate calling of germline and somatic variants. Nat Methods Nature Publishing Group 15(8):591–594

Kitzman JO (2016) Haplotypes drop by drop. Nat Biotechnol Nature Publishing Group 34(3):296–298

Knouse KA, Wu J, Whittaker CA, Amon A (2014) Single cell sequencing reveals low levels of aneuploidy across mammalian tissues. Proc Natl Acad Sci USA 11(37):13409–13414

Koboldt DC, Zhang Q, Larson DE, Shen D, McLellan MD, Lin L, Miller CA, Mardis ER, Ding L, Wilson RK (2012) VarScan 2: somatic mutation and copy number alteration discovery in cancer by exome sequencing. Genome Res 22(3):568–576. PMCID: PMC3290792

Korbel JO, Urban AE, Affourtit JP, Godwin B, Grubert F, Simons JF, Kim PM, Palejev D, Carriero NJ, Du L, Taillon BE, Chen Z, Tanzer A, Saunders ACE, Chi J, Yang F, Carter NP, Hurles ME, Weissman SM, Harkins TT, Gerstein MB, Egholm M, Snyder M (2007) Paired-end mapping reveals extensive structural variation in the human genome. Science 318(5849):420–426

Korbel JO, Abyzov A, Mu XJ, Carriero N, Cayting P, Zhang Z, Snyder M, Gerstein MB (2009) PEMer: a computational framework with simulation-based error models for inferring genomic structural variants from massive paired-end sequencing data. Genome Biol 10(2):R23. PMCID: PMC2688268

Kuijk E, Blokzijl F, Jager M, Besselink N, Boymans S, Chuva de Sousa Lopes SM (2019) van Boxtel R, Cuppen E. Early divergence of mutational processes in human fetal tissues. Sci Adv. American Association for the Advancement of Science 5(5):eaaw1271. PMCID: PMC6541467

Kunkel TA (2004) DNA replication fidelity. J Biol Chem 279(17):16895–16898

Laurie CC, Laurie CA, Rice K, Doheny KF, Zelnick LR, McHugh CP, Ling H, Hetrick KN, Pugh EW, Amos C, Wei Q, Wang L-E, Lee JE, Barnes KC, Hansel NN, Mathias R, Daley D, Beaty TH, Scott AF, Ruczinski I, Scharpf RB, Bierut LJ, Hartz SM, Landi MT, Freedman ND, Goldin LR, Ginsburg D, Li J, Desch KC, Strom SS, Blot WJ, Signorello LB, Ingles SA, Chanock SJ, Berndt SI, Le Marchand L, Henderson BE, Monroe KR, Heit JA, de Andrade M, Armasu SM, Regnier C, Lowe WL, Hayes MG, Marazita ML, Feingold E, Murray JC, Melbye M, Feenstra B, Kang JH et al (2012) Detectable clonal mosaicism from birth to old age and its relationship to cancer. Nat Genet. 44(6):642–650. PMCID: PMC3366033

Lawrence MS, Stojanov P, Polak P, Kryukov GV, Cibulskis K, Sivachenko A, Carter SL, Stewart C, Mermel CH, Roberts SA, Kiezun A, Hammerman PS, McKenna A, Drier Y, Zou L, Ramos AH, Pugh TJ, Stransky N, Helman E, Kim J, Sougnez C, Ambrogio L, Nickerson E, Shefler E, Cortés ML, Auclair D, Saksena G, Voet D, Noble M, DiCara D, Lin P, Lichtenstein L, Heiman DI, Fennell T, Imielinski M, Hernandez B, Hodis E, Baca S, Dulak AM, Lohr J, Landau D-A, Wu CJ, Melendez-Zajgla J, Hidalgo-Miranda A, Koren A, McCarroll SA, Mora J, Lee RS, Crompton B, Onofrio R et al (2013) Mutational heterogeneity in cancer and the search for new cancer-associated genes. Nature 499(7457):214–218. PMCID: PMC3919509

Lee JJ-K, Park S, Park H, Kim S, Lee J, Lee J, Youk J, Yi K, An Y, Park IK, Kang CH, Chung DH, Kim T-M, Jeon YK, Hong D, Park PJ, Ju YS, Kim YT (2019) Tracing oncogene rearrangements in the mutational history of lung adenocarcinoma. Cell 177(7):1842–1857.e21

Lee-Six H, Øbro NF, Shepherd MS, Grossmann S, Dawson K, Belmonte M, Osborne RJ, Huntly BJP, Martincorena I, Anderson E, O'Neill L, Stratton MR, Laurenti E, Green AR, Kent DG, Campbell PJ (2018) Population dynamics of normal human blood inferred from somatic mutations. Nature. Nature Publishing Group 561(7724):473–478. PMCID: PMC6163040

Leung ML, Wang Y, Waters J, Navin NE (2015) SNES: single nucleus exome sequencing. Genome Biol. BioMed Central 16(1):55. PMCID: PMC4373516

Livet J, Weissman TA, Kang H, Draft RW, Lu J, Bennis RA, Sanes JR, Lichtman JW (2007) Transgenic strategies for combinatorial expression of fluorescent proteins in the nervous system. Nature. Nature Publishing Group 450(7166):56–62

Lodato MA, Woodworth MB, Lee S, Evrony GD, Mehta BK, Karger A, Lee S, Chittenden TW, D'Gama AM, Cai X, Luquette LJ, Lee E, Park PJ, Walsh CA (2015) Somatic mutation in single human neurons tracks developmental and transcriptional history. Science. American Association for the Advancement of Science 350(6256):94–98. PMCID: PMC4664477

Lodato MA, Rodin RE, Bohrson CL, Coulter ME, Barton AR, Kwon M, Sherman MA, Vitzthum CM, Luquette LJ, Yandava CN, Yang P, Chittenden TW, Hatem NE, Ryu SC, Woodworth MB, Park PJ, Walsh CA (2018) Aging and neurodegeneration are associated with increased mutations in single human neurons. Science 359(6375):555–559. PMCID: PMC5831169

Loh P-R, Genovese G, Handsaker RE, Finucane HK, Reshef YA, Palamara PF, Birmann BM, Talkowski ME, Bakhoum SF, McCarroll SA, Price AL (2018) Insights into clonal haematopoiesis from 8,342 mosaic chromosomal alterations. Nature. Nature Publishing Group 559(7714):350–355. PMCID: PMC6054542

Marcy Y, Ishoey T, Lasken RS, Stockwell TB, Walenz BP, Halpern AL, Beeson KY, Goldberg SMD, Quake SR (2007) Nanoliter reactors improve multiple displacement amplification of genomes from single cells. PLoS Genet 3(9):e155

Maretty L, Jensen JM, Petersen B, Sibbesen JA, Liu S, Villesen P, Skov L, Belling K, Theil Have C, Izarzugaza JMG, Grosjean M, Bork-Jensen J, Grove J, Als TD, Huang S, Chang Y, Xu R, Ye W, Rao J, Guo X, Sun J, Cao H, Ye C, van Beusekom J, Espeseth T, Flindt E, Friborg RM, Halager AE, Le Hellard S, Hultman CM, Lescai F, Li S, Lund O, Løngren P, Mailund T, Matey-Hernandez ML, Mors O, Pedersen CNS, Sicheritz-Pontén T, Sullivan P, Syed A, Westergaard D, Yadav R, Li N, Xu X, Hansen T, Krogh A, Bolund L, Sørensen TIA, Pedersen O et al (2017) Sequencing and de novo assembly of 150 genomes from Denmark as a population reference. Nature. Nature Publishing Group 548(7665):87–91

Martincorena I, Roshan A, Gerstung M, Ellis P, Van Loo P, McLaren S, Wedge DC, Fullam A, Alexandrov LB, Tubio JM, Stebbings L, Menzies A, Widaa S, Stratton MR, Jones PH, Campbell PJ (2015) Tumor evolution. High burden and pervasive positive selection of somatic mutations in normal human skin. Science. American Association for the Advancement of Science 348(6237):880–886. PMCID: PMC4471149

Matevossian A, Akbarian S (2008) Neuronal nuclei isolation from human postmortem brain tissue. J Vis Exp (20):e914–4. PMCID: PMC3233860

McConnell MJ, Lindberg MR, Brennand KJ, Piper JC, Voet T, Cowing-Zitron C, Shumilina S, Lasken RS, Vermeesch JR, Hall IM, Gage FH (2013) Mosaic copy number variation in human neurons. Science 342(6158):632–637. PMCID: PMC3975283

McKenna A, Hanna M, Banks E, Sivachenko A, Cibulskis K, Kernytsky A, Garimella K, Altshuler D, Gabriel S, Daly M, DePristo MA (2010) The Genome Analysis Toolkit: a MapReduce framework for analyzing next-generation DNA sequencing data. Genome Res 20(9):1297–1303. PMCID: PMC2928508

Medvedev P, Stanciu M, Brudno M (2009) Computational methods for discovering structural variation with next-generation sequencing. Nat Methods 6(11 Suppl):S13–S20

Michaelson JJ, Shi Y, Gujral M, Zheng H, Malhotra D, Jin X, Jian M, Liu G, Greer D, Bhandari A, Wu W, Corominas R, Peoples A, Koren A, Gore A, Kang S, Lin GN, Estabillo J, Gadomski T, Singh B, Zhang K, Akshoomoff N, Corsello C, McCarroll S, Iakoucheva LM, Li Y, Wang J, Sebat J (2012) Whole-genome sequencing in autism identifies hot spots for de novo germline mutation. Cell 151(7):1431–1442. PMCID: PMC3712641

Milholland B, Auton A, Suh Y, Vijg J (2015) Age-related somatic mutations in the cancer genome. Oncotarget 6(28):24627–24635. PMCID: PMC4694783

Milholland B, Dong X, Zhang L, Hao X, Suh Y, Vijg J (2017) Differences between germline and somatic mutation rates in humans and mice. Nat Commun 8:15183. PMCID: PMC5436103

Navin N, Kendall J, Troge J, Andrews P, Rodgers L, McIndoo J, Cook K, Stepansky A, Levy D, Esposito D, Muthuswamy L, Krasnitz A, McCombie WR, Hicks J, Wigler M (2011) Tumour evolution inferred by single-cell sequencing. Nature 472(7341):90–94

O'Huallachain M, Karczewski KJ, Weissman SM, Urban AE, Snyder MP (2012) Extensive genetic variation in somatic human tissues. Proc Natl Acad Sci USA 109(44):18018–18023. PMCID: PMC3497787

Pan X, Urban AE, Palejev D, Schulz V, Grubert F, Hu Y, Snyder M, Weissman SM (2008) A procedure for highly specific, sensitive, and unbiased whole-genome amplification. Proc Natl Acad Sci USA 105(40):15499–15504. PMCID: PMC2563063

Picher ÁJ, Budeus B, Wafzig O, Krüger C, García-Gómez S, Martínez-Jiménez MI, Díaz-Talavera A, Weber D, Blanco L, Schneider A (2016) TruePrime is a novel method for whole-genome amplification from single cells based on TthPrimPol. Nat Commun 7:13296. PMCID: PMC5141293

Plusa B, Hadjantonakis A-K, Gray D, Piotrowska-Nitsche K, Jedrusik A, Papaioannou VE, Glover DM, Zernicka-Goetz M (2005) The first cleavage of the mouse zygote predicts the blastocyst axis. Nature Nature Publishing Group 434(7031):391–395

Podolskiy DI, Lobanov AV, Kryukov GV, Gladyshev VN (2016) Analysis of cancer genomes reveals basic features of human aging and its role in cancer development. Nat Commun. Nature Publishing Group 7:12157. PMCID: PMC4990632

Poduri A, Evrony GD, Cai X, Elhosary PC, Beroukhim R, Lehtinen MK, Hills LB, Heinzen EL, Hill A, Hill RS, Barry BJ, Bourgeois BFD, Riviello JJ, Barkovich AJ, Black PM, Ligon KL, Walsh CA (2012) Somatic activation of AKT3 causes hemispheric developmental brain malformations. Neuron 74(1):41–48. PMCID: PMC3460551

Rahbari R, Wuster A, Lindsay SJ, Hardwick RJ, Alexandrov LB, Turki Al S, Dominiczak A, Morris A, Porteous D, Smith B, Stratton MR (2016) UK10K Consortium, Hurles ME. Timing, rates and spectra of human germline mutation. Nat Genet 48(2):126–133. PMCID: PMC4731925

Ramsey MJ, Moore DH, Briner JF, Lee DA, Olsen LA, Senft JR, Tucker JD (1995) The effects of age and lifestyle factors on the accumulation of cytogenetic damage as measured by chromosome painting. Mutat Res 338(1–6):95–106

Saini N, Roberts SA, Klimczak LJ, Chan K, Grimm SA, Dai S, Fargo DC, Boyer JC, Kaufmann WK, Taylor JA, Lee E, Cortes-Ciriano I, Park PJ, Schurman SH, Malc EP, Mieczkowski PA, Gordenin DA (2016) The impact of environmental and endogenous damage on somatic mutation load in human skin fibroblasts. Taylor M, editor. PLoS Genet. Public Library of Science 12(10):e1006385. PMCID: PMC5082821

Shapiro E, Biezuner T, Linnarsson S (2013) Single-cell sequencing-based technologies will revolutionize whole-organism science. Nat Rev Genet Nature Publishing Group 14(9):618–630

Shen J-C, Rideout WM III, Jones PA (1994) The rate of hydrolytic deamination of 5-methylcytosine in double-stranded DNA. Nucleic Acids Res 22(6):972–976

Snippert HJ, van der Flier LG, Sato T, van Es JH, van den Born M, Kroon-Veenboer C, Barker N, Klein AM, van Rheenen J, Simons BD, Clevers H (2010) Intestinal crypt homeostasis results from neutral competition between symmetrically dividing Lgr5 stem cells. Cell 143(1):134–144

Solyom S, Kazazian HH (2012) Mobile elements in the human genome: implications for disease. Genome Med. BioMed Central 4(2):12. PMCID: PMC3392758

Spemann H, Evolution HMDGA (1924) Über induktion von Embryonalanlagen durch Implantation artfremder Organisatoren. Springer

Sturtevant AH (1929) The claret mutant type of Drosophila simulans: a study of chromosome elimination and of cell-lineage. Z Wiss Zool 135:323–356

Sulston JE, Schierenberg E, White JG, Thomson JN (1983) The embryonic cell lineage of the nematode Caenorhabditis elegans. Dev Biol 100(1):64–119

Tam PP, Parameswaran M, Kinder SJ, Weinberger RP (1997) The allocation of epiblast cells to the embryonic heart and other mesodermal lineages: the role of ingression and tissue movement during gastrulation. Development 124(9):1631–1642

Telenius H, Carter NP, Bebb CE, Nordenskjöld M, Ponder BA, Tunnacliffe A (1992) Degenerate oligonucleotide-primed PCR: general amplification of target DNA by a single degenerate primer. Genomics 13(3):718–725

Tomasetti C, Vogelstein B, Parmigiani G (2013) Half or more of the somatic mutations in cancers of self-renewing tissues originate prior to tumor initiation. Proc Natl Acad Sci USA (6):110, 1999–2004. PMCID: PMC3568331

Turner DL, Cepko CL (1987) A common progenitor for neurons and glia persists in rat retina late in development. Nature. Nature Publishing Group 328(6126):131–136

Upton KR, Gerhardt DJ, Jesuadian JS, Richardson SR, Sánchez-Luque FJ, Bodea GO, Ewing AD, Salvador-Palomeque C, van der Knaap MS, Brennan PM, Vanderver A, Faulkner GJ (2015) Ubiquitous L1 mosaicism in hippocampal neurons. Cell 161(2):228–239. PMCID: PMC4398972

Vogt W (1929) Gestaltungsanalyse am Amphibienkeim mit Örtlicher Vitalfärbung: II. Teil. Gastrulation und Mesodermbildung bei Urodelen und Anuren. Wilhelm Roux Arch Entwickl Mech Org 120(1):384–706

Wala JA, Bandopadhayay P, Greenwald NF, O'Rourke R, Sharpe T, Stewart C, Schumacher S, Li Y, Weischenfeldt J, Yao X, Nusbaum C, Campbell P, Getz G, Meyerson M, Zhang C-Z, Imielinski M, Beroukhim R (2018) SvABA: genome-wide detection of structural variants and indels by local assembly. Genome Res. Cold Spring Harbor Lab 28(4):581–591. PMCID: PMC5880247

Wald H (1936) Cytologic studies on the abnormal development of the eggs of the claret mutant type of Drosophila Simulans. Genetics. Genetics Society of America 21(3):264–281. PMCID: PMC1208673

Wang J, Fan HC, Behr B, Quake SR (2012) Genome-wide single-cell analysis of recombination activity and de novo mutation rates in human sperm. Cell 150(2):402–412

Wenger AM, Peluso P, Rowell WJ, Chang P-C, Hall RJ, Concepcion GT, Ebler J, Fungtammasan A, Kolesnikov A, Olson ND, Töpfer A, Alonge M, Mahmoud M, Qian Y, Chin C-S, Phillippy AM, Schatz MC, Myers G, DePristo MA, Ruan J, Marschall T, Sedlazeck FJ, Zook JM, Li H, Koren S, Carroll A, Rank DR, Hunkapiller MW (2019) Accurate circular consensus long-read sequencing improves variant detection and assembly of a human genome. Nat Biotechnol 74(11):5463

Wilson EB (1892) The cell-lineage of Nereis. A contribution to the cytogeny of the annelid body. J Morphol. John Wiley & Sons, Ltd 6(3):361–480

Yaniv K, Isogai S, Castranova D, Dye L, Hitomi J, Weinstein BM (2006) Live imaging of lymphatic development in the zebrafish. Nat Med Nature Publishing Group 12(6):711–716

Yao S, Sukonnik T, Kean T, Bharadwaj RR, Pasceri P, Ellis J (2004) Retrovirus silencing, variegation, extinction, and memory are controlled by a dynamic interplay of multiple epigenetic modifications. Mol Ther 10(1):27–36

Yuen RK, Merico D, Bookman M, Howe J, Thiruvahindrapuram B, Patel RV, Whitney J, Deflaux N, Bingham J, Wang Z, Pellecchia G, Buchanan JA, Walker S, Marshall CR, Uddin M, Zarrei M, Deneault E, D'Abate L, Chan AJS, Koyanagi S, Paton T, Pereira SL, Hoang N, Engchuan W, Higginbotham EJ, Ho K, Lamoureux S, Li W, MacDonald JR, Nalpathamkalam T, Sung WWL, Tsoi FJ, Wei J, Xu L, Tasse A-M, Kirby E, Van Etten W, Twigger S, Roberts W, Drmic I, Jilderda S, Modi BM, Kellam B, Szego M, Cytrynbaum C, Weksberg R, Zwaigenbaum L, Woodbury-Smith M, Brian J, Senman L et al (2017) Whole genome sequencing resource identifies 18 new candidate genes for autism spectrum disorder. Nat Neurosci 20(4):602–611. PMCID: PMC5501701

Yurov YB, Iourov IY, Vorsanova SG, Liehr T, Kolotii AD, Kutsev SI, Pellestor F, Beresheva AK, Demidova IA, Kravets VS, Monakhov VV, Soloviev IV (2007) Aneuploidy and confined chromosomal mosaicism in the developing human brain. PLoS ONE 2(6):e558. PMCID: PMC1891435

Zalokar M (1976) Division and migration of nuclei during early embryogenesis of Drosophila melanogaster. J Microsc Biol Cell 25:97–106

Zhang C-Z, Adalsteinsson VA, Francis J, Cornils H, Jung J, Maire C, Ligon KL, Meyerson M, Love JC (2015) Calibrating genomic and allelic coverage bias in single-cell sequencing. Nat Commun. Nature Publishing Group 6(1):6822–6810. PMCID: PMC4922254

Zhou T, Benda C, Duzinger S, Huang Y, Li X, Li Y, Guo X, Cao G, Chen S, Hao L, Chan Y-C, Ng K-M, Ho JC, Wieser M, Wu J, Redl H, Tse H-F, Grillari J, Grillari-Voglauer R, Pei D, Esteban MA (2011) Generation of induced pluripotent stem cells from urine. J Am Soc Nephrol 22(7):1221–1228. PMCID: PMC3137570

Zhou T, Benda C, Dunzinger S, Huang Y, Ho JC, Yang J, Wang Y, Zhang Y, Zhuang Q, Li Y, Bao X, Tse H-F, Grillari J, Grillari-Voglauer R, Pei D, Esteban MA (2012) Generation of human induced pluripotent stem cells from urine samples. Nat Protoc 7(12):2080–2089

Zinyk DL, Mercer EH, Harris E, Anderson DJ, Joyner AL (1998) Fate mapping of the mouse midbrain-hindbrain constriction using a site-specific recombination system. Curr Biol 8(11):665–668

Zong C, Lu S, Chapman AR, Xie XS (2012) Genome-wide detection of single-nucleotide and copy-number variations of a single human cell. Science 338(6114):1622–1626

Chapter 4
Interphase Chromosomes of the Human Brain

Yuri B. Yurov, Svetlana G. Vorsanova, and Ivan Y. Iourov

Abstract Molecular neurocytogenetic (neurocytogenomic) studies have shown the human brain to demonstrate somatic genome variability (mosaic aneuploidy, sub-chromosomal rearrangements). Chromosomal mosaicism and instability rates vary during ontogeny in the human brain: dramatic increase of the rates in the early brain development follows by a significant decrease in the postnatal period. It is highly likely that rates of mosaicism and instability increase in the aging brain. Alternatively, chromosome-specific instability (aneuploidy and interphase chromosome breaks) and increased levels of chromosomal mosaicism confined to the brain are associated with a wide spectrum of neurodevelopmental and neurodegenerative diseases. Neurocytogenetic/neurocytogenomic analyses may provide further insights into genome organization at the chromosomal level in cells of such a high-functioning system as the human brain. Here, we review studies of interphase chromosomes in the human brain. In this instance, the role of molecular neurocytogenetics and neurocytogenomics in current genetics, genomics, and cell biology of the human brain is discussed.

Y. B. Yurov · S. G. Vorsanova
Mental Health Research Center, Moscow, Russia

Veltischev Research and Clinical Institute for Pediatrics of the Pirogov Russian National Research Medical University, Ministry of Health of Russian Federation, Moscow, Russia

I. Y. Iourov (✉)
Mental Health Research Center, Moscow, Russia

Veltischev Research and Clinical Institute for Pediatrics of the Pirogov Russian National Research Medical University, Ministry of Health of Russian Federation, Moscow, Russia

Medical Genetics Department of Russian Medical Academy of Continuous Postgraduate Education, Moscow, Russia

Introduction

The availability of interphase molecular cytogenetic techniques (e.g., fluorescence in situ hybridization (FISH) with chromosome- and site-specific DNA probes) has made possible to analyze chromosomes in almost all cellular populations in humans (Soloviev et al. 1995; Yurov et al. 1996, 2013; Vorsanova et al. 2010c; Hu et al. 2020). Neural chromosomes have been found to demonstrate high rates of variations manifesting as aneuploidy (gain/loss of chromosomes in a cell), which has been hypothesized to mediate neuronal diversity and brain diseases. Currently, chromosomal variation in the human brain has shown to represent a mechanism for a variety of neurodegenerative and psychiatric diseases (Yurov et al. 2001, 2018b; Iourov et al. 2006c; Kingsbury et al. 2006; Arendt et al. 2009; Jourdon et al. 2020). Actually, one can distinguish two main directions of studying interphase chromosomes in the human brain: (I) analysis of numerical and structural chromosomal changes (i.e., aneuploidy, structural abnormalities, copy number variations (CNV), chromosome instability, etc.) and (II) uncovering genome organization at the chromosomal level. The former has been the focus of numerous molecular neurocytogenetic and neurocytogenomic studies, whereas the latter is likely to become a purpose of further neurocytogenetic research.

In the present chapter, we review the latest advances in studying chromosomes in the human brain at microscopic, submicroscopic, and molecular levels. Theoretical and practical issues of brain-specific cytogenomic analyses are considered.

Interphase Chromosomes and Brain Ontogeny: Natural Chromosomal Variations

The complexity, plasticity, and intercellular variability of the human brain are likely to be generated during early ontogenetic stages and to be mediated by genomic content of neural progenitor cells (Muotri and Gage 2006; Rohrback et al. 2018b). The developing mammalian brain is characterized by high levels of chromosomal variations affecting ~30% of cells (Rehen et al. 2001; Yurov et al. 2005, 2007a). More precisely, the developing human brain is demonstrated to possess 30–35% of aneuploid cells (1.25–1.45% per chromosome) revealed by methods based on fluorescence in situ hybridization (FISH). These are multiprobe FISH, quantitative FISH (QFISH), and interphase chromosome-specific multicolor banding (ICS-MCB) (Yurov et al. 2005, 2007a; Iourov et al. 2010a, 2019a) (Fig. 4.1). Additionally,

---

Fig. 4.1 (continued) (**d**) – chromosome 9, (**e**) – chromosome 16, and (**f**) – chromosome 18. (**g**) Interphase QFISH: (1) a nucleus with two signals for chromosomes 18 (relative intensities: 2058 and 1772 pixels), (2) a nucleus with one-paired signal mimics monosomy of chromosome 18 (relative intensity: 4012 pixels), (3) a nucleus with two signals for chromosomes 15 (relative intensities: 1562 and 1622 pixels), and (4) a nucleus with one signal showing monosomy of chromosome 15 (relative intensity: 1678 pixels). (From Yurov et al. 2007a, an open-access article distributed under the terms of the Creative Commons Attribution License)

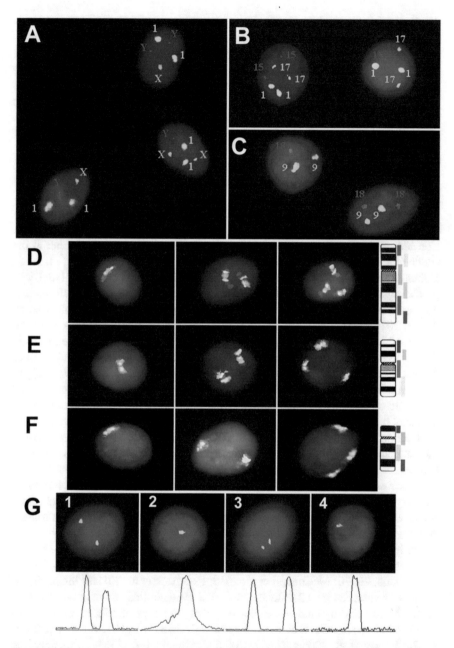

Fig. 4.1 Molecular cytogenetic analysis of aneuploidy in the fetal human brain. (**a**–**c**). Interphase FISH with chromosome-enumeration DNA probes: (**a**) two nuclei characterized by additional chromosomes Y and X and a normal nucleus; (**b**) a nucleus with monosomy of chromosome 15 and a normal nucleus; and (**c**) a nucleus with monosomy of chromosome 18 and a normal nucleus. (**d**–**g**) Interphase chromosome-specific MCB: nuclei with monosomy, disomy, trisomy, and G-banding ideograms with MCB color-code labeling of a chromosome (from left to right),

the developing human brain is the only embryonic tissue so far, which has demonstrated confined chromosomal mosaicism in contrast to confined placental mosaicism (Yurov et al. 2007a). At the subchromosomal level, similar progressive genomic changes are observed (i.e., high rates of brain-specific CNVs involving DNA sequences less than 1 Mb) in the developing human brain (McConnell et al. 2013; Rohrback et al. 2018a, b). At the sequence level per se, similar somatic genomic variations are unlikely to exist (Knouse et al. 2014; Muyas et al. 2020). Thus, (sub)chromosomal mosaicism and instability (aneuploidy) are hallmarks of the developing mammalian brain.

Taking into account a correlation between number of aneuploid cells (30–35%) and number of cells cleared by the programmed cell death (30–50%) in the developing brain, aneuploidization (progressive accumulation of aneuploid cells) is suggested as a mechanism for cell number regulation during early brain ontogeny (Iourov et al. 2006c; Muotri and Gage 2006; Yurov et al. 2010a; Fricker et al. 2018). Considering observations evaluating functional effects of aneuploidy either at the single cell level or at the tissular level (Iourov et al. 2008a; Dierssen et al. 2009; Hultén et al. 2013), mitotic catastrophe (a cascade of abnormal mitotic cell divisions producing aneuploidization) has been proposed as a mechanisms for cell number decreases in the developing brain because of aneuploid cell death (Iourov et al. 2006d, 2019d; Yurov et al. 2007a; Fricker et al. 2018). This hypothesis has been supported by studying chromosomal mosaicism in embryonic and extraembryonic tissues, which has shown that this mosaicism type is able to cause prenatal death or spontaneous abortions (Vorsanova et al. 2005, 2010a). Since aneuploidy is likely to have an adverse effect on cellular homeostasis, an alteration to the clearance of aneuploid cells during prenatal period may result in high rates of aneuploidy in the postnatal human brain, mediating neuropsychiatric and neurodegenerative diseases or childhood brain cancer (Iourov et al. 2006c, 2009c, 2019d; Kingsbury et al. 2006; McConnell et al. 2017; Yurov et al. 2018a, b, 2019b). On the other hand, aneuploidy may represent a mechanism for neuronal diversity in the unaffected human brain inasmuch as aneuploid neural cells are functionally active and integrated into brain circuitry (Kingsbury et al. 2005). To gain further insights into the role of chromosomal variation in the human brain in later ontogeny, one has to study interphase chromosome in the childhood and adult human brain.

During the prenatal period, rates of chromosomal and subchromosomal changes or instability decrease to 10% or lower (Yurov et al. 2005, 2018b, 2019b; Iourov et al. 2006a, 2009b; McConnell et al. 2013; Rohrback et al. 2018a). Interestingly, the way of variation in cell numbers mediated by aneuploidization in the developmental brain and programmed cell death is likely to be specific for humans in contrast to other vertebrates studied in this context (Rehen et al. 2001; Yurov et al. 2005, 2007a; Iourov et al. 2006c; Zupanc 2009; Rohrback et al. 2018a). Probably, the functional uniqueness of the human brain is achieved by such a kind of selective pressure at cellular/chromosomal level (Iourov et al. 2012, 2019d). Additionally, intercellular differences between DNA content (~250 Mb) in the adult human brain have been reported (Westra et al. 2008, 2010). The variability of the chromosomal numbers (aneuploidy) allowed to hypothesize that aneuploidy rates may be higher in late ontogeny. In other words, aneuploidization may be a mechanism for brain aging (Iourov et al. 2008a; Yurov et al. 2009b, 2010a, b; Faggioli et al. 2011). However, there is no consensus on the matter. Thus, a number of studies report increased rates of aneuploidy in the aged brain (Fischer et al. 2012; Andriani et al. 2017), whereas other reports do not (Van den Bos et al. 2016; Shepherd et al. 2018). The lack of consensus is more likely to be a result of technological differences between these reports. Single-cell sequencing studies report low rates of genomic changes in moderate cell numbers (~100 cell analyzed with the highest resolution possible) (Knouse et al. 2014; Van den Bos et al. 2016; Rohrback et al. 2018a), whereas molecular cytogenetic studies report high rates of chromosomal variations in large cell populations (reviewed by Iourov et al. 2012; Yurov et al. 2018b, 2019b). One can propose that combination of sequence-based single-cell techniques and molecular cytogenetic (cytogenomic) methods may solve the problem.

The devastating effect of chromosomal abnormalities (aneuploidy and structural aberrations) suggests that these genomic variations are able to produce functional and structural alterations to the human brain. The confinement of aneuploidy and other types of chromosomal variations (instability) to the central nervous system has been systematically associated with brain diseases (Yurov et al. 2001, 2018b; Iourov et al. 2006c, d, 2013; Tiganov et al. 2012; McConnell et al. 2017; Leija-Salazar et al. 2018; Iourov 2019; Potter et al. 2019; Heng 2020). It is highly likely that each form of brain pathology is linked to a specific type of brain-specific genomic alterations.

Interphase Chromosomes in the Diseased Brain

Chromosomal variations cause functional brain alterations in a wide spectrum of psychiatric and neurological diseases (DeLisi et al. 1994; Iourov et al. 2008b; Vorsanova et al. 2010d; Graham et al. 2019; Potter et al. 2019). Somatic genome variations at chromosomal and subchromosomal levels are repeatedly associated with neurodevelopmental, neurodegenerative, and/or psychiatric disorders (Iourov et al. 2008b, 2010b, 2019d; Smith et al. 2010; Paquola et al. 2017; Vorsanova et al. 2017; Graham et al. 2019). Chromosomal abnormalities and instability confined to the brain have been reported in schizophrenia and neurodegenerative diseases. Several neuropsychiatric diseases (e.g., autism and epilepsy) are also hypothesized to be associated with neurocytogenetic and neurocytogenomic variations.

The first report on two cases of mosaic aneuploidy (trisomy X and 18) in the schizophrenia brain (Yurov et al. 2001) has formed the basis for further neurocytogenomic studies of the diseased brain. As a result, several schizophrenia cases have been additionally associated with chromosome-1-specific instability and gonosomal instability, which are almost exclusively manifested as aneuploidy (Yurov et al. 2008, 2016, 2018a). Brain-specific structural chromosomal abnormalities (microdeletions) and CNV have been also found in a number of schizophrenia cases (Kim et al. 2014; Sakai et al. 2015). These data allow suggesting that a number of schizophrenia cases are the result of chromosomal abnormalities and/or instability in the diseased brain (Yurov et al. 2018a, b). Further molecular neurocytogenetic (neurocytogenomic) studies would certainly shed light on the involvement of "neurochromosomal variation" in schizophrenia and would likely to define the exact proportion of schizophrenia cases associated with neural aneuploidy, structural chromosome aberrations and chromosomal/genomic instability.

Somatic mosaic aneuploidy is one of the commonest types of genomic variations in autistic individuals inasmuch as ~10% of autistic males are likely to exhibit low-level 47,XXY/46,XY mosaicism (Yurov et al. 2007b). More importantly, gonosomal mosaicism is common in autistic individuals and their relatives. Several familial cases of behavioral abnormalities co-segregating with X chromosome aneuploidy and chromosomal instability have been reported (Vorsanova et al. 2007, 2010b). These data have been used for theoretical explanation of the male-to-female ratio in autism (Iourov et al. 2008c). Additionally, the neurocytogenetic hypothesis of autism (i.e., a proportion of autism cases may be associated with chromosome abnormalities and instability confined to the brain) has been recently described using systems biology methodology (Vorsanova et al. 2017). Our preliminary studies have demonstrated a possible involvement of brain-specific chromosome instability (chromothripsis) and aneuploidy in pathogenic cascades associated with autistic behavior (Iourov et al. 2017a). In the behavioral context, one has to mention studies suggesting that genome/chromosome instability probably shapes behavior in individuals suffering from neurodevelopmental diseases (Vorsanova et al. 2018) and gulf war illness (Liu et al. 2018). However, direct evaluation of interphase chromosomes in the autistic brain is still in process.

Somatic aneuploidy and other types of chromosome instability have been found to mediate neurodegeneration (Iourov et al. 2009a; Leija-Salazar et al. 2018; Shepherd et al. 2018; Yurov et al. 2019a). The Alzheimer's disease brain has been systematically shown to exhibit genome/chromosome instability and related phenomena (i.e., abnormal cell cycle entry, endomitosis, replication stress, abnormal DNA damage response, and micronuclei in mitotic tissues) (Herrup and Yang 2007; Mosch et al. 2007; Iourov et al. 2011; Yurov et al. 2011, 2019a; Arendt 2012; Bajic et al. 2015; Coppedè and Migliore 2015; Hou et al. 2017; Lin et al. 2020; Nudelman et al. 2019). Taking into account neurological parallels between Alzheimer's disease and Down syndrome or trisomy of chromosome 21 (Snyder et al. 2020), Professor Huntington Potter's group has proposed that brain-specific copy number changes of either whole chromosome 21 or chromosome 21 region containing *APP* gene are able to mediate neurodegeneration in Alzheimer's disease (Granic et al. 2010; Potter et al. 2019). Actually, chromosome 21-psecific instability in the diseased brain is one of the most probable mechanisms for Alzheimer's disease (Iourov et al. 2009b). Additionally, genes mutated in rare familial cases of the diseases are involved in processes granting proper chromosome segregation during the cell division (Boeras et al. 2008; Granic et al. 2010). Similarly, altered chromosome segregation induced by LDL/cholesterol seems to contribute to Alzheimer's disease as well as to Niemann-Pick C1 and atherosclerosis (Granic and Potter 2013). Moreover, X chromosome aneuploidy (X chromosome loss) — a cytogenetic biomarker of human aging — has been reported to have higher rates in the Alzheimer's disease brain as to the unaffected brain (Yurov et al. 2014) (Fig. 4.2). Selective cell death of aneuploid neurons (i.e., aneuploidy causes neuron death as it is the case in the developmental brain) has been reported to hallmark the neurodegeneration in the Alzheimer's disease brain (Arendt et al. 2010). Abnormal DNA damage response resulting in chromosome/genome instability is likely to result in neurodegeneration in the Alzheimer's disease brain (neural cells with aneuploidy or structurally altered chromosomes produced by DNA damage are susceptible to programmed cell death)

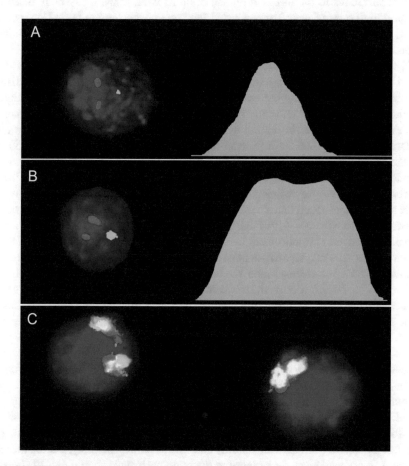

Fig. 4.2 Molecular neurocytogenetic analyses of the AD brain. (**a**) Multiprobe (two-probe) and quantitative FISH using DNA probes for chromosomes 1 (two red signals/D1Z1) and X (one green signal/DXZ1; relative intensity is 2120 pixels) demonstrating true X chromosome monosomy; (**b**) multiprobe (two-probe) and quantitative FISH using DNA probes for chromosomes 1 (two red signals/D1Z1) and X (one green signal/DXZ1; relative intensity is 4800 pixels) demonstrating overlapping of two X chromosome signals, but not a chromosome loss; (**c**) ICS-MCB with a probe set for chromosome X showing one nucleus bearing two chromosomes X and another nucleus bearing single chromosome X. (From Yurov et al. 2014, an open-access article distributed under the terms of the Creative Commons Attribution License)

(Fielder et al. 2017; Lin et al. 2020). Finally, Alzheimer's disease has been associated with subchromosomal instability (e.g., nonspecific CNVs) involving the *APP* gene (Kaeser and Chun 2020). In total, chromosome instability, including aneuploidy, represents an element of the Alzheimer's disease pathogenic cascade (Iourov et al. 2011; Yurov et al. 2019a). To link observations on aneuploidy/chromosome instability, abortive cell cycle, DNA damage, replication stress, and *APP*, a hypothesis depicted by Fig. 4.3 has been proposed.

Non-Alzheimer's disease neurodegeneration has been associated with chromosomal variations in the diseased human brain as well. Thus, Lewy body diseases exhibit high rates of neural aneuploidy in the neurodegenerating brain (Yang et al. 2015). *MAPT* mutations that lead to mitotic defects, neuronal aneuploidy and extensive apoptosis are likely to cause frontotemporal lobar degeneration (Caneus et al. 2018). Subchromosomal instability involving α-synuclein (*SNCA*) has been associated with Parkinson's disease and multiple system atrophy (Mokretar et al. 2018). Probably, the most intriguing example of a neurodegenerative disease associated with brain-specific chromosome instability is ataxia-telangiectasia, an autosomal recessive chromosome instability syndrome caused by *ATM* gene mutations and characterized by cerebellar degeneration (Iourov et al. 2007b; Potter et al. 2019). In fact, neurodegeneration caused by chromosome instability has been firstly demonstrated during the molecular cytogenetic analysis of the ataxia-telangiectasia brain (previously, chromosome instability has been suggested to be almost exclusive mechanism for cancer) (Iourov et al. 2009a, b). The ataxia-telangiectasia brain demonstrates chromosome-14 instability (interphase chromosomal breaks and additional rearranged chromosomes) in ~40% of cells in the degenerating cerebellum (Iourov et al. 2009a). These data have been used as a basis for potential therapeutic strategies for neurodegeneration mediated by chromosome (genome) instability (Yurov et al. 2009a; Iourov et al. 2019b). There are striking differences between cancerous chromosome instability and neurodegenerative chromosome instability. The differences are as follows: **Cancer**: Cancer-susceptibility mutations interact with environment producing genome and chromosome instabilities. These processes lead to clonal evolution and, thereby, malignancy. **Neurodegeneration**: Chromosome instability and abnormalities are present in a significant proportion of cells, and genetic-environment interactions trigger progressive neuronal cell loss (neurodegeneration) by natural selection and/or programmed cell death (Iourov et al. 2013; Yurov et al. 2019a). Schematically, this model is shown by Fig. 4.4.

In the previous version of the book (Yurov et al. 2013), we proposed a hypothesis describing the role of neural aneuploidy and chromosome instability. During the last 7 years, more evidences for supporting the hypothesis have been provided (Iourov et al. 2014, 2019a, b, d; Yurov et al. 2014, 2018a, b, 2019a, b; Bajic et al. 2015; Andriani et al. 2017; McConnell et al. 2017; Vorsanova et al. 2017, 2020; Leija-Salazar et al. 2018; Rohrback et al. 2018b; Shepherd et al. 2018; Graham et al. 2019; Iourov 2019; Potter et al. 2019; Jourdon et al. 2020). Accordingly, we would like to reproduce schematically the hypothesis (Fig. 4.5).

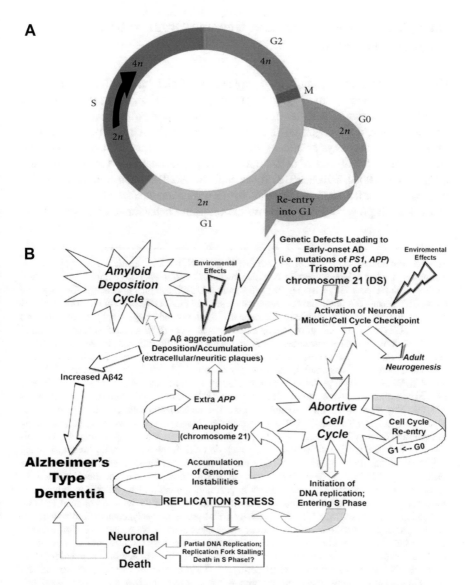

Fig. 4.3 (**a**) Simplified schematic presentation of the cell cycle theory of AD. Quiescent neuronal cells (G0 phase) demonstrate the cell cycle reactivation by either endogenous or environmental mitogenic stimuli followed by reentry into the G1 phase. The G0/G1 phase transition is critical for a postmitotic neuron and potentially causes neuronal cell death. During G1 phase, diploid neurons (chromosomal complement: 2 N; number of chromosomes: 46; DNA content: 2C) demonstrate G1-specific cell cycle markers (cyclin D and CDK4/6 complex, cyclin E, and CDK2 complex) which are involved in the regulation of G1 phase progression. Cells successfully passing G1 enter the S phase (phase of DNA replication). During the S phase, CDK2/cyclin E should be silenced to repress additional round of replication of genomic DNA. Protein markers of the S phase are A-type cyclins (cyclin A/CDK2 complex). This complex is essential for proper completion of S phase and transition from S to G2 phase. DNA content of cells during S phase changes from 2C to

(continued)

Interphase Chromosomes and Genome Organization in the Human Brain

Nuclear genome organization in interphase is crucial for regulating chromatin remodeling, genome activity (transcription), genome safeguarding (DNA damage response, proper chromosome segregation, mitotic checkpoint, etc.), DNA repair and replication, and programmed cell death (for details, see Chaps. 1, 2, and 9). Previously, we have systematically indicated the importance of neurocytogenetic analysis of chromosome organization in interphase nuclei of the human brain (Iourov et al. 2006c, 2010a, 2012; Yurov et al. 2013, 2018b). Unfortunately, no significant progress has been, as yet, made in this field. Nonetheless, we have attempted to list known properties of interphase chromosome behavior in the human brain

Fig. 4.3 (continued) 4C (chromosome number is still 2 N, but DNA content after replication is tetraploid). During G2 phase, cyclin A is degraded, and cyclin B/CDC2 complex (protein biomarker of late S/early G2 phases) is formed. Cyclin B/CDC2 complex is essential for triggering mitosis. Neuronal cells in G2 phase demonstrate tetraploid (4 N) DNA content or, more precisely, possess a nucleus with 46 replicated chromosomes. Chromosomal complement (genomic content) of cells in G2 consists of one set of 46 duplicated chromosomes (DNA content: 4 N or 4C; diploid nucleus with replicated chromosomes; for more details see, [20]), each having two chromatids— "mitotic" tetraploidy. It is to note that true constitutional polyploidy is a term used to describe cell containing more than two homologous sets of chromosomes (4 N or 92 chromosomes, DNA content: 4C). We suggest that postmitotic neurons are able to replicate DNA but are not able to make a G2/M transition and divide into two daughter cells. (**b**) The DNA replication stress hypothesis of AD. Interplay between essential elements of the AD-type dementia pathogenetic cascade is proposed. The genetic influences (PSEN or APP mutations, trisomy 21, APOE4 genotype), metabolic changes, and environmental factors affecting neuronal homeostasis in the aging brain lead to activation of neuronal proliferation. Mitogens, which do exist in the human brain (neuronal cells), induce additional stimuli of extensive adult neurogenesis in the hippocampus. In the AD brain, such events would lead to increased hippocampal neurogenesis. A side effect could be that these mitogenic stimuli activate cell cycle reentry in postmitotic neurons. The latter is a pathological activation of neuronal cell cycle, including reentry into G1 and S phases and initiation of DNA replication. Neurons showing protein markers of G2/M phase, probably, contain chromosome set of 23 duplicated chromosome pairs with unseparated chromatids (DNA content, 4C; chromosome complement, 2 N) and become tetraploid in a sense of DNA content (4C). According to the commonly accepted theory of neuronal cell cycle reentry and death, some neuronal populations complete the DNA synthesis but are arrested during the G2/M transition. Therefore, neuronal death occurs in G2 phase. Alternatively, one can propose that a large proportion of activated postmitotic neurons in the AD brain are unable to pass properly the S phase. This would lead to accumulation of genomic and chromosomal instabilities throughout ontogeny (DNA breaks, aneuploidy). In addition, replication-induced DNA damages would lead to fork stalling, incomplete or inefficient DNA replication, together designated as replication stress. Replication stress may be considered the leading cause of neuronal cell death due to processing into S phase or accumulation of genetic instabilities, which together constitute an important element of the AD pathogenetic cascade. According to the present hypothesis, the possibility to link the two main pathways of AD arises from the introduction of accumulation of genomic instabilities associated with DNA replication stress, which is able to produce as neuronal cell death (replicative cell death) as chromosomal aneuploidy due to natural selection in neural cell populations probably causing extra *APP* in the diseased brain. (From Yurov et al. 2011, an open-access article distributed under the terms of the Creative Commons Attribution License)

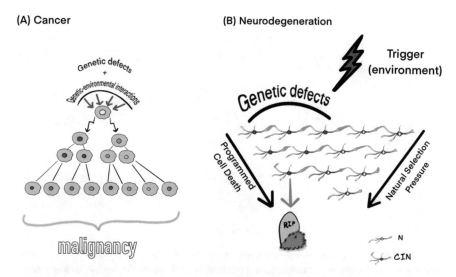

Fig. 4.4 Theoretical model for CIN mediating (**a**) cancer and (**b**) neurodegeneration. (**a**) Genetic defects and genetic-environmental interactions may cause chromosomal/genomic changes, which produce CIN; alternatively, cell populations may adapt to aneuploidy and CIN evolving to a cell population with a fitness advantage. Cells affected by CIN and tolerating deteriorating effects of CIN on cellular homeostasis are able to evolve clonally to produce malignancy. (**b**) CIN/somatic mosaicism affecting a significant proportion of cells interacting with environmental triggers may result into progressive neuronal cell loss (neurodegeneration) under natural selection pressure and through the programmed cell death (N, normal neurons; CIN, neuronal cell affected by CIN). The model is based on the observations of CIN in the neurodegenerating brain and cancers. (From Yurov et al. 2019a, an open-access article distributed under the terms of the Creative Commons Attribution License)

along with molecular cytogenetic FISH-based techniques, which are used for the analysis.

To perform a successful study of chromosomal arrangement in interphase, one has to be aware about the spatial preservation of interphase nuclei during tissue/cell suspension preparation for molecular cytogenetic analysis. Although brain cell preparation for molecular neurocytogenetic analysis requires specific procedures, it does provide an opportunity to preserve interphase nuclei of the human brain (Iourov et al. 2006b; Yurov et al. 2017b). Pairing of homologous chromosomes (chromosomal associations/locus associations) is common in the postnatal human brain (Iourov et al. 2005, 2017b; Yurov et al. 2017b). To make accurate scoring of the associations, QFISH may be applied (Iourov et al. 2005; Iourov 2017). Finally, functional complexity and structural variability of neural cell populations lead to requirement of studying integral interphase chromosomes at molecular resolutions in a "band-by-band" manner. This technical opportunity is offered by interphase chromosome-specific multicolor banding (ICS-MCB) (Iourov et al. 2006a, 2007a). An example of ICS-MCB is shown by Fig. 4.6. Nuclear genome organization at the chromosomal level may be a mechanism for brain diseases (Iourov 2012; Yurov

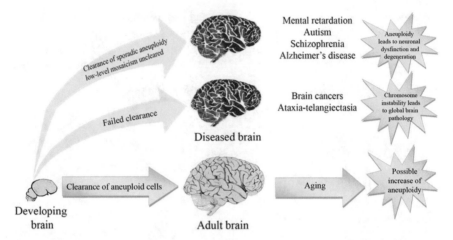

Fig. 4.5 Schematic representation of the hypothesis on the role of aneuploidy in normal CNS development and aging as well as in pathogenesis of brain diseases. During the normal prenatal brain development, developmental chromosome instability is cleared leading to three-time decrease of aneuploidy rates. Brain aging is likely to be associated with slight increase of aneuploidy. Total failure of clearance of developmental chromosome instability would lead to the persistence as observed in chromosome instability syndromes with brain dysfunction (ataxia-telangiectasia) and brain cancers. Clearance may not affect low-level chromosomal mosaicism confined to the developing brain, which is extremely frequent among human fetuses. In such cases, the postnatal brain exhibits low-level chromosome-specific mosaic aneuploidy. The latter is shown to be associated with diseases of neuronal dysfunction and degeneration (mental retardation, autism, schizophrenia, Alzheimer's disease). (From Yurov et al. 2013 (previous edition of the book — Figure 4.9), reproduced with permission of Springer Nature in the format reuse in a book/textbook via Copyright Clearance Center)

et al. 2013). However, there are no, as yet, studies attempting to correlate specific nuclear chromosome organization in neural cells and central nervous system dysfunction.

Conclusion

The present chapter is dedicated to behavior and variation of interphase chromosomes in the human brain. Aneuploidy and other types of chromosome instability are mechanisms for neuronal diversity and brain diseases. As repeatedly noted before, brain-oriented interphase chromosome (neurocytogenetic and neurocytogenomic) analysis brings new insights to neuroscience, human genomics, and molecular medicine.

Molecular (neuro)cytogenetic and (neuro)cytogenomic studies seem to benefit from bioinformatics approaches based on network- or pathway-based analysis, i.e., systems biology methodology (Yurov et al. 2017a, b). Actually, pathway-based classification of human diseases is considered the most promising way to unravel

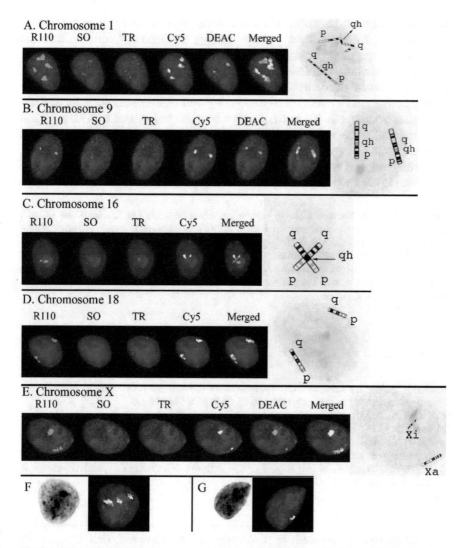

Fig. 4.6 FISH using MCB probes on interphase nuclei of the human brain. (**a**): FISH with MCB probe for chromosome 1. R110 signals correspond to 1p32.3Yp36.3 and 1q32Yq43. SO (Spectrum Orange) signals Y 1p13Yq21 including constitutive heterochromatin (1qh). TR (Texas Red) signals Y 1p31.1Yp33 and 1q21.3Yq31. Cy5 signals Y 1p13.1Yp22.3 and 1q32Yq43. DEAC signals Y 1q21.3Yq31. Note the upper chromosome 1 is folded around 1qh and bent in the proximal part of the q-arm. (**b**): FISH with MCB probe for chromosome 9. R110 signals correspond to 9p13Yq13 including constitutive heterochromatin (9qh). SO (Spectrum Orange) signals Y 9p21Yp24 and 9q32Yq34. TR (Texas Red) signals Y 9q22.2Yq34.1. Cy5 signals Y 9p13Yp23. DEAC signals Y 9q13Yq22.2. (**c**): FISH with MCB probe for chromosome 16. R110 signals correspond to 16p11.1Yp13.1 SO (Spectrum Orange) signals Y 16p13.3Yp21. TR (Texas Red) signals Y 16q11.1Yq21 including constitutive heterochromatin (16qh). Cy5 signals Y 16q21Yq24. Note the single Texas Red signal instead of two; this implies that 16qh regions of two homologous chromosomes 16 are overlapped. Therefore, somatic pairing of two homologous chromosomes 16 by 16qh region should be suspected. (**d**): FISH with MCB probe for chromosome 18. R110 signals

(continued)

complex relationship between molecular/cellular processes and phenotypes (Iourov et al. 2019b). We suggest that systems biology methodology considered in the molecular cytogenomic context is able to provide new information about interphase chromosomes in the human brain (Yurov et al. 2017a, b; Iourov et al. 2019c). These approaches toward the definition of molecular basis of human brain diseases have been already found successful: (i) uncovering molecular mechanisms for somatic mosaicism (Iourov et al. 2015), (ii) genomic instability associated with neurological and psychiatric diseases (McConnell et al. 2017; Vorsanova et al. 2017), and (iii) molecular/cellular alterations causing brain dysfunction (Iourov et al. 2009b, 2019b, c). To this end, one has to conclude that interphase chromosome studies certainly contribute to our knowledge about the human central nervous system.

Acknowledgments We would like to express our gratitude to Dr. OS Kurinnaia and Dr. MA Zelenova for help in chapter preparation. Professors SG Vorsanova and IY Iourov are partially supported by RFBR and CITMA according to the research project No. 18-515-34005. Prof. IY Iourov's lab is supported by the Government Assignment of the Russian Ministry of Science and Higher Education, Assignment no. AAAA-A19-119040490101-6. Prof. SG Vorsanova's lab is supported by the Government Assignment of the Russian Ministry of Health, Assignment no. AAAA-A18-118051590122-7.

References

Andriani GA, Vijg J, Montagna C (2017) Mechanisms and consequences of aneuploidy and chromosome instability in the aging brain. Mech Ageing Dev 161:19–36
Arendt T (2012) Cell cycle activation and aneuploid neurons in Alzheimer's disease. Mol Neurobiol 46(1):125–135
Arendt T, Mosch B, Morawski M (2009) Neuronal aneuploidy in health and disease: a cytomic approach to understand the molecular individuality of neurons. Int J Mol Sci 10(4):1609–1627
Arendt T, Brückner MK, Mosch B et al (2010) Selective cell death of hyperploid neurons in Alzheimer's disease. Am J Pathol 177:15–20

Fig. 4.6 (continued) correspond to 18p11.2Yq12.2. SO (Spectrum Orange) signals Y 18p11.2Yp11.3. TR (Texas Red) signals Y 18q22Yq23. Cy5 signals Y 18q11.2Yq21.3. (**e**): FISH with MCB probe for chromosome X. R110 signals correspond to Xp21.3Yp22.3 and Xq25Yq28. SO (Spectrum Orange) signals Y Xp11.22Yp22.1 and Xq25Yq28. TR (Texas Red) signals Y Xq12Yq21.1. Cy5 signals Y Xq21.1Yq26. DEAC signals Y Xp11.3Yq13. Note the upper chromosome X appears as a white condensed spot (merged image). Since facultative heterochromatin, a feature of X chromosome inactivation, should appear as a highly condensed structure, the upper X chromosome was assumed to be inactivated one (Xi) in contrast to the active X chromosome (Xa) appearing as a slightly diffused structure. (**f**): Example of a trisomic nucleus (trisomy of chromosome 9); left side, Y black-and-white picture of DAPI-counterstained nucleus, and right side, Y merged MCB true color picture showing the presence of three chromosomes 9 in this nucleus. (**g**): Example of a monosomic nucleus (monosomy of chromosome 18); left side, Y black-and-white picture of DAPI-counterstained nucleus, and right side, Y merged MCB true color picture showing the presence of one chromosome 18 in this nucleus. (From Iourov et al. 2006a, reproduced with permission of Springer Nature in the format reuse in a book/textbook via Copyright Clearance Center)

Bajic V, Spremo-Potparevic B, Zivkovic L et al (2015) Cohesion and the aneuploid phenotype in Alzheimer's disease: a tale of genome instability. Neurosci Biobehav Rev 55:365–374

Boeras DI, Granic A, Padmanabhan J et al (2008) Alzheimer's presenilin 1 causes chromosome missegregation and aneuploidy. Neurobiol Aging 29:319–328

Caneus J, Granic A, Rademakers R et al (2018) Mitotic defects lead to neuronal aneuploidy and apoptosis in frontotemporal lobar degeneration caused by *MAPT* mutations. Mol Biol Cell 29:575–586

Coppedè F, Migliore L (2015) DNA damage in neurodegenerative diseases. Mutat Res 776:84–97

DeLisi LE, Friedrich U, Wahlstrom J et al (1994) Schizophrenia and sex chromosome anomalies. Schizophr Bull 20(3):495–505

Dierssen M, Herault Y, Estivill X (2009) Aneuploidy: from a physiological mechanism of variance to down syndrome. Physiol Rev 89:887–920

Faggioli F, Vijg J, Montagna C (2011) Chromosomal aneuploidy in the aging brain. Mech Ageing Dev 132(8–9):429–436

Fielder E, von Zglinicki T, Jurk D (2017) The DNA damage response in neurons: die by apoptosis or survive in a senescence-like state? J Alzheimers Dis 60:S107–S131

Fischer HG, Morawski M, Brückner MK et al (2012) Changes in neuronal DNA content variation in the human brain during aging. Aging Cell 11(4):628–633

Fricker M, Tolkovsky AM, Borutaite V et al (2018) Neuronal cell death. Physiol Rev 98:813–880

Graham EJ, Vermeulen M, Vardarajan B et al (2019) Somatic mosaicism of sex chromosomes in the blood and brain. Brain Res 1721:146345

Granic A, Potter H (2013) Mitotic spindle defects and chromosome mis-segregation induced by LDL/cholesterol-implications for Niemann-Pick C1, Alzheimer's disease, and atherosclerosis. PLoS One 8:e60718

Granic A, Padmanabhan J, Norden M et al (2010) Alzheimer Abeta peptide induces chromosome mis-segregation and aneuploidy, including trisomy 21: requirement for au and APP. Mol Biol Cell 21(4):511–520

Heng HH (2020) New data collection priority: focusing on genome-based bioinformation. Res Res Biomed 6(1):5–8

Herrup K, Yang Y (2007) Cell cycle regulation in the postmitotic neuron: oxymoron or new biology? Nat Rev Neurosci 8:368–378

Hou Y, Song H, Croteau DL et al (2017) Genome instability in Alzheimer disease. Mech Ageing Dev 161:83–94

Hu Q, Maurais EG, Ly P (2020) Cellular and genomic approaches for exploring structural chromosomal rearrangements. Chromosom Res 28(1):19–30

Hultén MA, Jonasson J, Iwarsson E et al (2013) Trisomy 21 mosaicism: we may all have a touch of down syndrome. Cytogenet Genome Res 139(3):189–192

Iourov IY (2012) To see an interphase chromosome or: how a disease can be associated with specific nuclear genome organization. BioDiscovery 4:e8932

Iourov IY (2017) Quantitative fluorescence *in situ* hybridization (QFISH). Methods Mol Biol 1541:143–149

Iourov IY (2019) Cytopostgenomics: what is it and how does it work? Curr Genomics 20(2):77–78

Iourov IY, Soloviev IV, Vorsanova SG et al (2005) An approach for quantitative assessment of fluorescence *in situ* hybridization (FISH) signals for applied human molecular cytogenetics. J HistochemCytochem 53:401–408

Iourov IY, Liehr T, Vorsanova SG et al (2006a) Visualization of interphase chromosomes in postmitotic cells of the human brain by multicolour banding (MCB). Chromosom Res 14(3):223–229

Iourov IY, Vorsanova SG, Pellestor F et al (2006b) Brain tissue preparations for chromosomal PRINS labeling. Methods Mol Biol 334:123–132

Iourov IY, Vorsanova SG, Yurov YB (2006c) Chromosomal variation in mammalian neuronal cells: known facts and attractive hypotheses. Int Rev Cytol 249:143–191

Iourov IY, Vorsanova SG, Yurov YB (2006d) Intercellular genomic (chromosomal) variations resulting in somatic mosaicism: mechanisms and consequences. Curr Genomics 7:435–446

Iourov IY, Liehr T, Vorsanova SG et al (2007a) Interphase chromosome-specific multicolor band-
ing (ICS-MCB): a new tool for analysis of interphase chromosomes in their integrity. Biomol
Eng 24(4):415–417
Iourov IY, Vorsanova SG, Yurov YB (2007b) Ataxia telangiectasia paradox can be explained by
chromosome instability at the subtissue level. Med Hypotheses 68:716
Iourov IY, Vorsanova SG, Yurov YB (2008a) Chromosomal mosaicism goes global. Mol
Cytogenet 1:26
Iourov IY, Vorsanova SG, Yurov YB (2008b) Molecular cytogenetics and cytogenomics of brain
diseases. Curr Genomics 9(7):452–465
Iourov IY, Yurov YB, Vorsanova SG (2008c) Mosaic X chromosome aneuploidy can help to
explain the male-to-female ratio in autism. Med Hypotheses 70:456
Iourov IY, Vorsanova SG, Liehr T et al (2009a) Increased chromosome instability dramatically dis-
rupts neural genome integrity and mediates cerebellar degeneration in the ataxia-telangiectasia
brain. Hum Mol Genet 18(14):2656–2669
Iourov IY, Vorsanova SG, Liehr T et al (2009b) Aneuploidy in the normal, Alzheimer's disease
and ataxia-telangiectasia brain: differential expression and pathological meaning. Neurobiol
Dis 34(2):212–220
Iourov IY, Vorsanova SG, Yurov YB (2009c) Developmental neural chromosome instability as a
possible cause of childhood brain cancers. Med Hypotheses 72:615–616
Iourov IY, Vorsanova SG, Solov'ev IV et al (2010a) Methods of molecular cytogenetics for study-
ing interphase chromosome in human brain cells. Russ J Genet 46(9):1039–1041
Iourov IY, Vorsanova SG, Yurov YB (2010b) Somatic genome variations in health and disease.
Curr Genomics 11:387–396
Iourov IY, Vorsanova SG, Yurov YB (2011) Genomic landscape of the Alzheimer's disease brain:
chromosome instability – aneuploidy, but not tetraploidy – mediates neurodegeneration.
Neurodegener Dis 8:35–37
Iourov IY, Vorsanova SG, Yurov YB (2012) Single cell genomics of the brain: focus on neuronal
diversity and neuropsychiatric diseases. Curr Genomics 13(6):477–488
Iourov IY, Vorsanova SG, Yurov YB (2013) Somatic cell genomics of brain disorders: a new oppor-
tunity to clarify genetic-environmental interactions. Cytogenet Genome Res 139(3):181–188
Iourov IY, Vorsanova SG, Liehr T et al (2014) Mosaike im Gehirn des Menschen. Diagnostische
Relevanz in der Zukunft? Med Genet 26(3):342–345
Iourov IY, Vorsanova SG, Zelenova MA et al (2015) Genomic copy number variation affecting
genes involved in the cell cycle pathway: implications for somatic mosaicism. Int J Genomics
2015:757680
Iourov IY, Vorsanova SG, Liehr T et al (2017a) Chromothripsis as a mechanism driving genomic
instability mediating brain diseases. Mol Cytogenet 10(1):20(O2)
Iourov IY, Vorsanova SG, Yurov YB (2017b) Interphase FISH for detection of chromosomal mosaicism. In: Liehr T (ed) Fluorescence in situ hybridization (FISH) — application guide (springer
protocols handbooks), 2nd edn. Springer-Verlag, Berlin, Heidelberg, pp 361–372
Iourov IY, Liehr T, Vorsanova SG et al (2019a) The applicability of interphase chromosome-
specific multicolor banding (ICS-MCB) for studying neurodevelopmental and neurodegenera-
tive disorders. Res Results Biomedicine 5(3):4–9
Iourov IY, Vorsanova SG, Yurov YB (2019b) Pathway-based classification of genetic diseases. Mol
Cytogenet 12:4
Iourov IY, Vorsanova SG, Yurov YB (2019c) The variome concept: focus on CNVariome. Mol
Cytogenet 12:52
Iourov IY, Vorsanova SG, Yurov YB et al (2019d) Ontogenetic and pathogenetic views on somatic
chromosomal mosaicism. Genes (Basel) 10(5):E379
Jourdon A, Fasching L, Scuderi S et al (2020) The role of somatic mosaicism in brain disease. Curr
Opin Genet Dev 65:84–90
Kaeser GE, Chun J (2020) Mosaic somatic gene recombination as a potentially unifying hypoth-
esis for Alzheimer's disease. Front Genet 11:390

Kim J, Shin JY, Kim JI et al (2014) Somatic deletions implicated in functional diversity of brain cells of individuals with schizophrenia and unaffected controls. Sci Rep 4:3807

Kingsbury MA, Friedman B, McConnell MJ et al (2005) Aneuploid neurons are functionally active and integrated into brain circuitry. Proc Natl Acad Sci U S A 102:6143–6147

Kingsbury MA, Yung YC, Peterson SE et al (2006) Aneuploidy in the normal and diseased brain. Cell Mol Life Sci 63:2626–2641

Knouse KA, Wu J, Whittaker CA et al (2014) Single cell sequencing reveals low levels of aneuploidy across mammalian tissues. Proc Natl Acad Sci U S A 111:13409–13414

Leija-Salazar M, Piette C, Proukakis C (2018) Somatic mutations in neurodegeneration. Neuropathol Appl Neurobiol 4:267–285

Lin X, Kapoor A, Gu Y et al (2020) Contributions of DNA damage to Alzheimer's disease. Int J Mol Sci 21:1666

Liu G, Ye CJ, Chowdhury SK et al (2018) Detecting chromosome condensation defects in gulf war illness patients. Curr Genomics 19:200–206

McConnell MJ, Lindberg MR, Brennand KJ et al (2013) Mosaic copy number variation in human neurons. Science 342:632–637

McConnell MJ, Moran JV, Abyzov A et al (2017) Intersection of diverse neuronal genomes and neuropsychiatric disease: The Brain Somatic Mosaicism Network. Science 356:eaal1641

Mokretar K, Pease D, Taanman JW et al (2018) Somatic copy number gains of α-synuclein (*SNCA*) in Parkinson's disease and multiple system atrophy brains. Brain 141:2419–2431

Mosch B, Morawski M, Mittag A et al (2007) Aneuploidy and DNA replication in the normal human brain and Alzheimer's disease. J Neurosci 27:6859–6867

Muotri AR, Gage FH (2006) Generation of neuronal variability and complexity. Nature 441:1087–1093

Muyas F, Zapata L, Guigó R et al (2020) The rate and spectrum of mosaic mutations during embryogenesis revealed by RNA sequencing of 49 tissues. Genome Med 12:49

Nudelman KNH, McDonald BC, Lahiri DK et al (2019) Biological hallmarks of cancer in Alzheimer's disease. Mol Neurobiol 56:7173–7187

Paquola ACM, Erwin JA, Gage FH (2017) Insights into the role of somatic mosaicism in the brain. Curr Opin Syst Biol 1:90–94

Potter H, Chial HJ, Caneus J et al (2019) Chromosome instability and mosaic aneuploidy in neurodegenerative and neurodevelopmental disorders. Front Genet 10:1092

Rehen SK, McConnell MJ, Kaushal D et al (2001) Chromosomal variation in neurons of the developing and adult mammalian nervous system. Proc Natl Acad Sci U S A 98:13361–13366

Rohrback S, April C, Kaper F et al (2018a) Submegabase copy number variations arise during cerebral cortical neurogenesis as revealed by single-cell whole-genome sequencing. Proc Natl Acad Sci U S A 115:10804–10809

Rohrback S, Siddoway B, Liu CS et al (2018b) Genomic mosaicism in the developing and adult brain. Dev Neurobiol 78:1026–1048

Sakai M, Watanabe Y, Someya T et al (2015) Assessment of copy number variations in the brain genome of schizophrenia patients. Mol Cytogenet 8:46

Shepherd CE, Yang Y, Halliday GM (2018) Region- and cell-specific aneuploidy in brain aging and neurodegeneration. Neuroscience 374:326–334

Smith CL, Bolton A, Nguyen G (2010) Genomic and epigenomic instability, fragile sites, schizophrenia and autism. Curr Genomics 11(6):447–469

Snyder HM, Bain LJ, Brickman AM et al (2020) Further understanding the connection between Alzheimer's disease and down syndrome. Alzheimers Dement 16(7):1065–1077

Soloviev IV, Yurov YB, Vorsanova SG et al (1995) Prenatal diagnosis of trisomy 21 using interphase fluorescence *in situ* hybridization of post-replicated cells with site-specific cosmid and cosmid contig probes. Prenat Diagn 15:237–248

Tiganov AS, Iurov IB, Vorsanova SG et al (2012) Genomic instability in the brain: etiology, pathogenesis and new biological markers of psychiatric disorders. Vestn Ross Akad Med Nauk 67(9):45–53

Van den Bos H, Spierings DC, Taudt AS et al (2016) Single-cell whole genome sequencing reveals no evidence for common aneuploidy in normal and Alzheimer's disease neurons. Genome Biol 17:116

Vorsanova SG, Kolotii AD, Iourov IY et al (2005) Evidence for high frequency of chromosomal mosaicism in spontaneous abortions revealed by interphase FISH analysis. J Histochem Cytochem 53(3):375–380

Vorsanova SG, Yurov IY, Demidova IA et al (2007) Variability in the heterochromatin regions of the chromosomes and chromosomal anomalies in children with autism: identification of genetic markers of autistic spectrum disorders. Neurosci Behav Physiol 37(6):553–558

Vorsanova SG, Iourov IY, Kolotii AD et al (2010a) Chromosomal mosaicism in spontaneous abortions: analysis of 650 cases. Rus J Genet 46:1197–1200

Vorsanova SG, Voinova VY, Yurov IY et al (2010b) Cytogenetic, molecular-cytogenetic, and clinical-genealogical studies of the mothers of children with autism: a search for familial genetic markers for autistic disorders. Neurosci Behav Physiol 40(7):745–756

Vorsanova SG, Yurov YB, Iourov IY (2010c) Human interphase chromosomes: a review of available molecular cytogenetic technologies. Mol Cytogenet 3:1

Vorsanova SG, Yurov YB, Soloviev IV et al (2010d) Molecular cytogenetic diagnosis and somatic genome variations. Curr Genomics 11(6):440–446

Vorsanova SG, Yurov YB, Iourov IY (2017) Neurogenomic pathway of autism spectrum disorders: linking germline and somatic mutations to genetic-environmental interactions. Curr Bioinforma 12(1):19–26

Vorsanova SG, Zelenova MA, Yurov YB et al (2018) Behavioral variability and somatic mosaicism: a cytogenomic hypothesis. Curr Genomics 19(3):158–162

Vorsanova SG, Yurov YB, Iourov IY (2020) Dynamic nature of somatic chromosomal mosaicism, genetic-environmental interactions and therapeutic opportunities in disease and aging. Mol Cytogenet 13:16

Westra JW, Peterson SE, Yung YC et al (2008) Aneuploid mosaicism in the developing and adult cerebellar cortex. J Comp Neurol 507:1944–1951

Westra JW, Rivera RR, Bushman DM et al (2010) Neuronal DNA content variation (DCV) with regional and individual differences in the human brain. J Comp Neurol 518:3981–4000

Yang Y, Shepherd C, Halliday G (2015) Aneuploidy in Lewy body diseases. Neurobiol Aging 36:1253–1260

Yurov YB, Soloviev IV, Vorsanova SG et al (1996) High resolution multicolor fluorescence in situ hybridization using cyanine and fluorescein dyes: rapid chromosome identification by directly fluorescently labeled alphoid DNA probes. Hum Genet 97(3):390–398

Yurov YB, Vostrikov VM, Vorsanova SG et al (2001) Multicolor fluorescent in situ hybridization on post-mortem brain in schizophrenia as an approach for identification of low-level chromosomal aneuploidy in neuropsychiatric diseases. Brain and Development 23(1):S186–S190

Yurov YB, Iourov IY, Monakhov VV et al (2005) The variation of aneuploidy frequency in the developing and adult human brain revealed by an interphase FISH study. J Histochem Cytochem 53(3):385–390

Yurov YB, Iourov IY, Vorsanova SG et al (2007a) Aneuploidy and confined chromosomal mosaicism in the developing human brain. PLoS One 2(6):e558

Yurov YB, Vorsanova SG, Iourov IY et al (2007b) Unexplained autism is frequently associated with low-level mosaic aneuploidy. J Med Genet 44(8):521–525

Yurov YB, Iourov IY, Vorsanova SG et al (2008) The schizophrenia brain exhibits low-level aneuploidy involving chromosome 1. Schizophr Res 98:139–147

Yurov YB, Iourov IY, Vorsanova SG (2009a) Neurodegeneration mediated by chromosome instability suggests changes in strategy for therapy development in ataxia-telangiectasia. Med Hypotheses 73:1075–1076

Yurov YB, Vorsanova SG, Iourov IY (2009b) GIN'n'CIN hypothesis of brain aging: deciphering the role of somatic genetic instabilities and neural aneuploidy during ontogeny. MolCytogenet 2:23

Yurov YB, Vorsanova SG, Iourov IY (2010a) Ontogenetic variation of the human genome. Curr Genomics 11(6):420–425

Yurov YB, Vorsanova SG, Solov'ev IV et al (2010b) Instability of chromosomes in human nerve cells (normal and with neuromental diseases). Russ J Genet 46(10):1194–1196

Yurov YB, Vorsanova SG, Iourov IY (2011) The DNA replication stress hypothesis of Alzheimer's disease. ScientificWorldJournal 11:2602–2612

Yurov YB, Vorsanova SG, Iourov IY (2013) Human interphase chromosomes — biomedical aspects. Springer New York, Heidelberg, Dordrecht, London

Yurov YB, Vorsanova SG, Liehr T et al (2014) X chromosome aneuploidy in the Alzheimer's disease brain. Mol Cytogenet 7(1):20

Yurov YB, Vorsanova SG, Demidova IA et al (2016) Genomic instability in the brain: chromosomal mosaicism in schizophrenia. Zh Nevrol Psikhiatr Im S Psychiatry Korsakova 116(11):86–91

Yurov YB, Vorsanova SG, Iourov IY (2017a) Network-based classification of molecular cytogenetic data. Curr Bioinforma 12(1):27–33

Yurov YB, Vorsanova SG, Soloviev IV et al (2017b) FISH-based assays for detecting genomic (chromosomal) mosaicism in human brain cells. NeuroMethods 131:27–41

Yurov YB, Vorsanova SG, Demidova IA et al (2018a) Mosaic brain aneuploidy in mental illnesses: an association of low-level post-zygotic aneuploidy with schizophrenia and comorbid psychiatric disorders. Curr Genomics 19(3):163–172

Yurov YB, Vorsanova SG, Iourov IY (2018b) Human molecular neurocytogenetics. Curr Genet Med Rep 6(4):155–164

Yurov YB, Vorsanova SG, Iourov IY (2019a) Chromosome instability in the neurodegenerating brain. Front Genet 10:892

Yurov YB, Vorsanova SG, Iourov IY (2019b) FISHing for unstable cellular genomes in the human brain. OBM Genetics 3(2):11

Zupanc GK (2009) Towards brain repair: insights from teleost fish. Semin Cell Dev Biol 20:683–690

Chapter 5
Senescence and the Genome

Joanna M. Bridger and Helen A. Foster

Abstract Cellular senescence is commonly initiated in response to replicative or cell stress pathways. Senescent cells remain in a state of permanent cell cycle arrest, and although being metabolically active, they exhibit distinct senescence phenotypes. Though cellular senescence may be beneficial in tumour suppression and wound healing, it is commonly associated with age-related diseases. There are various mechanisms and drivers that contribute to ageing, but it is becoming increasingly apparent that processes related to chromatin and the epigenome are also important. Indeed, three of the nine hallmarks of ageing are genome specific including genomic instability, epigenetic alterations and telomere attrition. With the advent of new technologies like DNA adenine methyltransferase identification and chromosome conformation capture, the features and complexity of the ageing genome are being revealed. This chapter will address key characteristics of interphase nuclei during cellular senescence including the spatio-temporal organisation of chromosomes, chromatin remodelling and epigenome changes.

The Senescence Phenotype

The term "cellular senescence" was originally coined by Hayflick in 1965 (Hayflick 1965). It was described as an important mechanism to suppress tumorigenicity. Senescence can be categorised into four types as shown in Fig. 5.1: (1) replicative senescence (RS) as a result of telomere dysfunction or shortening; (2) genotoxic stress-induced senescence due to endogenous stress, e.g. oxidative stress and severe or irreparable DNA damage; (3) oncogene-induced senescence (OIS) via the activation of aberrant signalling pathways caused by different mechanisms including

J. M. Bridger
Centre for Genome Engineering and Maintenance, Department of Life Sciences, Division of Biosciences, College of Health, Medicine and Life Sciences, Brunel University London, Uxbridge, UK

H. A. Foster (✉)
Department of Clinical, Pharmaceutical and Biological Science, School of Life and Medical Sciences, University of Hertfordshire, Hatfield, UK
e-mail: h.foster2@herts.ac.uk

© Springer Nature Switzerland AG 2020
I. Iourov et al. (eds.), *Human Interphase Chromosomes*,
https://doi.org/10.1007/978-3-030-62532-0_5

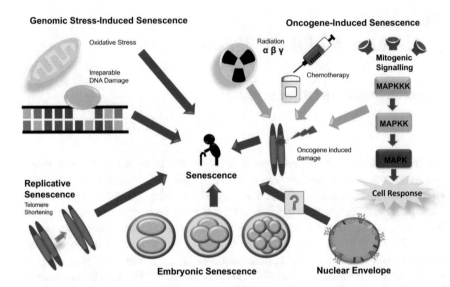

Fig. 5.1 The four main categories of senescence. Genomic stress-induced senescence can be induced via products from cellular metabolism, e.g. reactive oxygen species (ROS) produced by mitochondria or DNA damage due to errors in DNA replication, recombination or repair mechanisms. Activation of aberrant signalling pathways can lead to oncogene-induced senescence and may result from mitogenic signalling or genotoxic agents such as chemical mutagens and radioactivity. Embryonic senescence is important for developmentally regulated growth and patterning. Replicative senescence can lead to irreversible cell cycle arrest by telomere shortening or dysfunction

natural endogenous processes such as mitogenic signalling or oxidative respiration, physical or chemical insults encountered during life or therapeutic treatment such as irradiation or chemotherapy; and (4) embryonic-senescence which occurs in a developmentally regulated manner (Coppé et al. 2010; Munoz-Espin et al. 2013; Storer et al. 2013; Graziano and Gonzalo 2017). The characteristics of senescence may vary depending on the mechanism by which it was induced; for instance, a senescence-associated secretory phenotype (SASP) that secretes proinflammatory mediators is present with some forms of senescence, but not others (Coppé et al. 2010). Regardless of the type of senescence, they each share the characteristic arrest in cell proliferation (Coppé et al. 2010). This chapter will concentrate on how the genome and its behaviour are altered during senescence.

Organisation of Chromatin and the Epigenome During Ageing

DNA contains genetic information that, when expressed, ultimately codes for the synthesis of a range of proteins vital for the correct functioning of cells, tissues and the whole organism. The genome, when housed in cell nuclei, needs to be organised correctly so that this information can be safely conveyed during proliferation and cell division to daughter cells, be protected from damage and allow genes to be expressed or repressed depending on the protein requirements of the cell and differentiated tissue. The nucleosome is an octamer of histone proteins composed of two copies of histones H2A, H2B, H3 and H4. There are approximately 30 million nucleosomes within the genome, and DNA wraps around these nucleosome complexes to form chromatin (Xu and Liu 2019). Epigenetics involves heritable changes that alter the expression of genes but do not change the DNA sequence. These modifications can act directly on the DNA by adding methyl groups to cytosine; through post-translational modifications to histones including acetylation, methylation, phosphorylation, sumoylation or ubiquitination; or via non-coding RNAs such as microRNA (miRNA), Piwi-interacting RNA (piRNA) and small interfering RNA (siRNA) (Dupont et al. 2009; Wei et al. 2017). Deregulation of epigenetic mechanisms has been highlighted in disease aetiology and ageing. Epigenetic clocks predict the chronological age of individuals by studying the methylation status of cytosines in specific GC-rich regions of the genome, known as CpG islands (Horvath 2013). Mathematical algorithms are employed to determine DNA methylation levels (5-methylcytosine or 5mC) from sets of CpG islands to estimate the age of the DNA source (Horvath and Raj 2018). CpG islands are commonly found near promoter regions of the genome, are $\geq$0.5 kb long with a GC content of $\geq$55% and are generally unmethylated (Jeziorska et al. 2017). Global DNA hypomethylation, with hypermethylation of specific loci, is associated with physiological ageing (Gensous et al. 2017). Changes in DNA methylation have been demonstrated in a number of age-related diseases such as cancer (Xie et al. 2019), Parkinson's disease (Miranda-Morales et al. 2017; Navarro-Sánchez et al. 2018) and Alzheimer's disease (Levine et al. 2015) as well in cells derived from Hutchinson-Gilford progeria syndrome (HGPS) patients (Ehrlich 2019). However, epigenetic changes during ageing are complex. Although region-specific hypermethylation may be determined at specific CpG islands and gene loci, ageing is also associated with global hypomethylation across the genome (Gensous et al. 2017) and loss of heterochromatin (Goldman et al. 2004; Chandra et al. 2015).

The degree of chromatin compaction can vary in cells, with euchromatin being less compact and open in structure and heterochromatin being more condensed. Generally, euchromatin is rich in CpG islands, has a high GC content, is gene-dense and is associated with short interspersed elements (SINEs) and transcriptional activity (Medstrand et al. 2002; Elbarbary et al. 2016; Vanrobays 2017). Conversely, heterochromatin is AT-rich and gene-poor, associated with long interspersed elements (LINEs), and is inaccessible to transcription factors (Vanrobays 2017; Medstrand et al. 2002; Elbarbary et al. 2016). Epigenetically, histones in heterochromatin generally have methylated H3K9 and H3K27, whilst euchromatin has both acetylation and methylation of H3K4 and H3K36 (Ahringer and Gasser 2018). Heterochromatin can be further subdivided into constitutive heterochromatin and facultative heterochromatin. Constitutive heterochromatin is not transcribed, contains highly repetitive sequences and is H3K9 methylated to maintain a stable condensed state important for chromosome structure such as in centromeres and telomeres (Ahringer and Gasser 2018). Facultative heterochromatin is reversible and may adopt both open or compact conformations according to (1) spatial parameters, e.g. changes in nuclear localisation due to factors such as signalling; (2) temporal changes, e.g. within the cell cycle or development; or (3) heritable factors, e.g. chromosome X inactivation (Trojer and Reinberg 2007). Thus, euchromatin has the potential to be decondensed and express genes in certain tissues. Euchromatic and heterochromatic domains are established during embryogenesis and development to generate tissue-specific gene expression patterns (Villeponteau 1997). Commonly within interphase nuclei, heterochromatin is concentrated at the nuclear periphery, nucleoli, centromeres and telomeres (Goldman et al. 2002), whilst euchromatin is positioned within the nuclear interior (Romero-Bueno et al. 2019). However, ageing is associated with substantial changes in heterochromatin distribution and epigenetic modifications.

During ageing, altered histone modifications and the redistribution of heterochromatin is thought to be associated with changes in global gene expression and genomic instability. Whole-genome bisulfite sequencing (WGBS) and CpG DNA methylation microarrays have been used to examine the epigenetic profiles of samples derived from a newborn and centenarian (103-year-old) (Heyn et al. 2012). Overall, the centenarian sample had a lower DNA methylation content, with the most hypomethylated sequences in CpG-poor promoters and tissue-specific genes (Heyn et al. 2012). Interestingly, the methylation status in middle-aged adults showed an intermediate level of global DNA methylation, suggesting an accumulative change with advancing age (Heyn et al. 2012). This is not unique as loss of heterochromatin has also been linked to an ageing phenotype in model organisms including *Caenorhabditis elegans* and *Drosophila* (Haithcock et al. 2005; Larson et al. 2012; Maleszewska et al. 2016). Modifications to histones are made through histone-modifying enzymes including histone methyltransferases, histone demethylases, histone deacetylases and histone acetylases (Black et al. 2012). Therefore, changes in expression or activity of these enzymes may have a profound influence on the epigenetic landscape of the genome. This is observed in *Arabidopsis thaliana* whereby reduced transcription of methyltransferases and increased transcription of demethylases are associated with hypomethylation in ageing (Ogneva et al. 2016). Furthermore, mutations in a H3K4 methyltransferase in *C. elegans* and yeast have been shown to reduce longevity, whilst reduced levels of H3K36 demethylase increases lifespan (Sen et al. 2015; Ni et al. 2012).

Epigenetic changes during ageing and loss of heterochromatin also contribute to the derepression of previously silenced genes at those loci (Sturm et al. 2015). This can result in the activation and potential remobilisation of transposable elements (TEs) throughout the genome (Sturm et al. 2015). Given that nearly half of the human genome consists of TEs, this could lead to genomic instability if a TE were to relocate into a coding or regulatory sequence within the genome (Mills et al. 2006; de Koning et al. 2011; Sturm et al. 2015). Ultimately, the resulting DNA damage and instability may result in age-related diseases such as cancer (O'Donnell and Burns 2010), and there are data to demonstrate this mobility, enhancing senescence in humans (Baillie et al. 2011; De Cecco et al. 2013; Keyes 2013).

Nucleosome density has been shown to alter during ageing and is associated with a loss of histones (Hu et al. 2014; Song and Johnson 2018). Nucleosome density naturally varies across the genome with transcriptionally active regions having a lower density and more open chromatin and transcriptionally inactive regions being densely populated with nucleosomes (Boeger et al. 2003; Sidler et al. 2017). Loss of nucleosomes in yeast leads to an increase in transcriptional activity from previously repressed promoters and corresponds with extensive chromosomal alterations and elevation of DNA strand breaks (Hu et al. 2014). Changes in nucleosome density could be due to two mechanisms: (1) alterations in the activity of histone chaperones and (2) reduction in histone biogenesis within the cell (Booth and Brunet 2016). There is evidence that nucleosome assembly may be regulated by the histone chaperone ASF1 in both a DNA synthesis-dependent and DNA synthesis-independent manner along with other histone chaperones, chromatin assembly factor 1 (CAF-1) and histone repression A factor (HIRA) (Galvani et al. 2008). In yeast, loss of function of ASF1 may lead to aberrant heterochromatin formation and genomic instability (Tanae et al. 2012). Indeed, ASF1 expression decreases with increasing age in human cells (O'Sullivan et al. 2010). Here, the synthesis of histones in fibroblasts derived from an old individual was half that compared to those derived from a child (O'Sullivan et al. 2010). Histone biosynthesis was also altered in replicative senescent IMR90 and WI38 cells, leading to downregulation of the synthesis of histones H3 and H4 and post-translational modifications (O'Sullivan et al. 2010; Song and Johnson 2018). Thus, nucleosome density combined with changes in epigenetic post-translational modifications could be an important factor in the loss of heterochromatin observed in ageing.

Conversely, there are regions of the genome that become associated with heterochromatin during ageing (Tsurumi and Li 2012). Chromatin may be organised within senescence-associated heterochromatin foci (SAHF) (Morris et al. 2019; Lenain et al. 2017; Braig et al. 2005; Michaloglou et al. 2005; Haugstetter et al. 2010). SAHF share epigenetic features and characteristics commonly found in heterochromatin including late replicating DNA domains (Shah et al. 2013); epigenetic markers H3K9me3 and H3K27me3 (Chandra et al. 2012; Chandra et al. 2015; Chandra and Narita 2013); HP1 α, β and γ (Boumendil et al. 2019); heterochromatic proteins; histone variant macroH2A; and high-mobility group A (HMGA) proteins (Morris et al. 2019). SAHF structure encompasses a chromatin core that is compacted and enriched in H3K9m3 (a marker of constitutive heterochromatin) and an outer ring of chromatin containing H3K27me3 (a marker of facultative heterochromatin), which is protein rich but more relaxed (Lenain et al. 2017; Chandra et al. 2012; Sadaie et al. 2013). Over 90% of SAHF are commonly observed in cells that have undergone OIS (Chandra et al. 2015) with only a small proportion seen in replicative senescence in cultures (Chandra et al. 2015; Boumendil et al. 2019). SAHFs are not present in HGPS or senescent mouse cells, and it is unclear if they occur in vivo (Lazzerini Denchi et al. 2005; Shumaker et al. 2006; Scaffidi and Misteli 2006; Swanson et al. 2013). The formation of SAHF represses the expression of genes that are important for proliferation and the cell cycle such as cyclin A, proliferating nuclear antigen (PCNA), E2F target genes (Aird and

Zhang 2013) and cyclin D1 (Zhang et al. 2007; Park et al. 2018) and thus leads to senescence. Evidence suggests that SAHF may result from an increase in nuclear pore density during OIS, with the nucleoporin TPR having a vital role in inducing the formation of SAHF and their maintenance (Boumendil et al. 2019).

Epigenetic modifications and heterochromatin distribution are altered in premature ageing syndromes. The majority of premature ageing syndromes are caused by either mutations leading to alterations in the nuclear lamina and matrix proteins or via defects in DNA repair systems (Musich and Zou 2009; Tiwari and Wilson 3rd 2019). Hutchinson-Gilford progeria syndrome (HGPS) is a premature ageing disease caused by a mutated lamin A protein. Here, there is a reduction in H3K9me3 and HP1 and loss of peripheral heterochromatin (Shumaker et al. 2006; Scaffidi and Misteli 2006). Werner syndrome (WS) is another progeroid syndrome that similarly has a loss of H3K9me3. WS is caused by mutations within the Werner helicase (WRN). Interestingly, WRN has been shown to associate with the methyltransferase SUV39H1 and HP1α and thus may be important in regulating heterochromatin during ageing (Zhang et al. 2015; Wang et al. 2016). Mesenchymal stem cells with an induced WRN deficiency show altered heterochromatin distribution and global loss of associated epigenetic methylation of histone H3 (Zhang et al. 2015; Shumaker et al. 2006). Loss of peripheral heterochromatin adjacent to the nuclear envelope (NE) (Goldman et al. 2004; Zhang et al. 2015) and a reduction in H3K9me3 and H3K27me3 levels but increase in H4K27me3 have been shown in HGPS cultured cells (Shumaker et al. 2006; Scaffidi and Misteli 2006).

Nuclear Lamina and Nucleoskeleton

The nuclear lamina is located adjacent to the inner nuclear membrane (INM) and is composed of type V intermediate filaments proteins—lamins and lamina-associated proteins. Lamins are subdivided into the B-type lamins which are constitutively expressed within mammalian cells and A-type lamins that are developmentally regulated in differentiated cells. The nuclear lamina interacts with INM proteins, nuclear pore complexes and chromatin. It has a number of important roles including organising chromatin, involvement in DNA replication and gene expression and to support structurally the nucleus and its processes (Cau et al. 2014). The peripheral nuclear lamina is interconnected and part of a larger structural protein network known as the nuclear matrix (NM) (Cau et al. 2014) or nucleoskeleton. This structure is believed to be a filamentous meshwork of proteins (e.g. lamins A and C), DNA and RNA localised throughout the nucleoplasm that are resistant to high-salt treatment and nucleases during experiments. Similarly, the matrix structure is important for the structural integrity of nuclei and also supports gene expression, chromatin organisation, DNA replication and repair (Chattopadhyay and Pavithra 2007; Wilson and Coverley 2017; Bridger et al. 2014; Mehta et al. 2007; Elcock and Bridger 2008; Godwin et al. 2021). The NM interacts with chromatin typically via specialised AT-rich DNA sequences called scaffold/matrix attachment regions (S/

MARs) (Barboro et al. 2012) and helps maintain the compartmentalisation of the nucleus and higher-order chromatin organisation important for the spatio-temporal dynamics of the cell.

Laminopathies include progeroid syndromes linked to A-type lamin mutations. These are typically characterised as having nuclear envelope deformities (Cau et al. 2014), with blebbing, herniations, invaginations and altered nuclear shape. These are caused by mutations that influence the post-translational processing of proteins, ultimately leading to defective protein function. For instance, in HGPS, there is a cryptic splice site that leads to a truncated form of lamin A which is permanently bound to a farnesyl moiety, termed "progerin" (Gilbert and Swift 2019). The build-up of progerin at the INM is toxic, leading to altered nuclear envelope integrity and perturbed chromatin organisation (Chandra et al. 2015; Stephens et al. 2018; Bikkul et al. 2018). Although progerin is primarily associated with HGPS, it has been suggested that a progerin-dependent mechanism may lead to natural ageing (Scaffidi and Misteli 2006; McClintock et al. 2007; Ashapkin et al. 2019). Evidence acquired by reverse transcription polymerase chain reaction (RT-PCR) has shown that fibroblasts obtained from naturally aged individuals expressed progerin mRNA, albeit at a low frequency of less than 50-fold (Scaffidi and Misteli 2006). Progerin has also been detected in cell lines derived from skin biopsies that had undergone prolonged cell culture, particularly in cells derived from older individuals (McClintock et al. 2007). However, it should be noted that the levels were very low. Senescence is frequently accompanied with profound changes to the INM organisation and accompanying processes.

The nuclear lamina interacts with the genome directly through lamina-associated domains (LADs) and via lamin-binding partners. DNA adenine methyltransferase identification (DamID) technology has been used to extensively map LADs throughout the nucleus. This technique is used to identify binding sites between DNA and chromatin-binding proteins. For instance, combining a nuclear lamin protein (e.g. lamin B1) to a bacterial DNA adenine methyltransferase (Dam) will highlight areas of DNA that have been in contact with the nuclear lamins as they will undergo adenine methylation. As adenine methylation does not naturally occur in eukaryotes, it acts as a detectable marker. LADs are of fundamental importance in anchoring transcriptionally silent heterochromatin to the nuclear lamina and maintaining the three-dimensional spatial arrangement of chromosomes (van Steensel and Belmont 2017; Romero-Bueno et al. 2019). However, lamins can be found throughout the nucleoplasm (Bridger et al. 1993) and not just at the nuclear envelope, and thus, this should be taken into consideration. LADs are also heterogeneous between cell types (Peric-Hupkes et al. 2010; Meuleman et al. 2013) but are associated with lamin B1 and lamin B1 receptor (LBR), which anchor heterochromatin to the nuclear lamina (Lukasova et al. 2018). However, during cellular senescence, LADs become extensively redistributed (Lochs et al. 2019). Normally, after DNA replication, DNA methyltransferase DNMT1 restores the histone methylation pattern; however, this appears to fail during senescence (Lochs et al. 2019) leading to hypomethylation. This hypomethylation, combined with the loss of lamin B1, leads to heterochromatin dissociating from the nuclear lamina

(Lochs et al. 2019) and away from the nuclear periphery. This LAD rearrangement may also be associated with the accumulation of SAHF, which relocates heterochromatin to the nuclear interior (Lenain et al. 2017; Chandra et al. 2015).

Nucleolus

The nucleolus is also important for spatio-temporal regulation of the genome and is formed from chromosomes containing active nucleolar organiser regions (NORs) and other non-acrocentric but gene-rich chromosomes (van Koningsbruggen et al. 2010; Nemeth et al. 2010). Nucleoli are important in ribosome biogenesis and are initiated from the transcription of ribosomal RNA (rRNA) genes found in high copy number and arranged in tandem repeats within NORs (Bersaglieri and Santoro 2019). Genomic regions that are localised in close proximity to the nucleolus are termed nucleolus-associated domains (NADs) (Nemeth et al. 2010). Genome-wide mapping has demonstrated that NADs derived from HeLa, IMR90 and HT1080 human cell lines have a low gene density, low transcriptional levels, late-replicating loci and heterochromatin enriched with repressive histone modifications H4K20me3, H3K27me3 and H3K9me3 (van Koningsbruggen et al. 2010; Nemeth et al. 2010; Dillinger et al. 2017). During senescence, nucleoli may fuse and are associated with an increased size (Mehta et al. 2007). H3K9me3 modified heterochromatin localised at the nucleolus, is remodelled and is coupled with an observed dissociation of centromeric and pericentromeric satellite regions away from the nucleolus (Dillinger et al. 2017). Interestingly, mapping of NADs using Hi-C in senescent cells remains similar to that seen in proliferating cell lines although there are changes in sub-NADs association with nucleoli (subdomains smaller than 100 kb), which appear to correspond to transcriptional changes (Dillinger et al. 2017; Mehta et al. 2010).

Centromeres and Telomeres

Centromeres are heterochromatic regions and have satellite II and α-satellite repeat sequences that are normally constitutively repressed (De Cecco et al. 2013). However, in replicative senescent cells, the pericentric satellite has been shown to distend, and chromatin is reorganised becoming more accessible and hypomethylated (De Cecco et al. 2013; Cruickshanks et al. 2013). This centromere distension has been termed "senescence-associated distension of satellites" (SADS) and is associated with epigenetic modifications associated with early senescence (Criscione et al. 2016). Silencing of pericentric satellite DNA is helped and maintained by SIRT6, a histone deacetylase, which removes H3K18 acetylation in normal proliferative cells (Nagai et al. 2015; Tasselli et al. 2016). It is possible that SIRT6 depletion could lead to senescence (Tasselli et al. 2016; Nagai et al. 2015). Indeed, SIRT6 is an early factor sequestered to double-strand breaks, so prolonged recruitment to irreversibly damaged DNA associated with ageing may lead to depletion of SIRT6 at pericentric satellite DNA leading to the unravelling and SADS phenotype (Toiber et al. 2013; Nagai et al. 2015; Tasselli et al. 2016). SADS occurs as a common feature of senescence, irrespective of how senescence is induced and whether the p16 or p21 pathways are activated (Swanson et al. 2013). Unlike SAHF, SADS are found both during normal senescence and in progeria (Swanson et al. 2013).

Telomeres, and their associated shelterin protein complex, are located at the ends of linear chromosomes and have a protective role in preventing genome instability by shielding exposed ends of DNA. During replication, DNA polymerases are unable to completely replicate the telomere region of the lagging strand leading to shortening due to the progressive loss of telomere repeats. This has been termed the "end-replication problem." Consequently, the length of the telomeres shortens with each cell division leading to attrition. This has been extensively reported in ageing studies and is particularly pronounced within in vitro primary cells leading to a finite number of cell divisions or "replicative senescence" due to the shortened telomere lengths. The resulting exposure of the chromosome ends leads to the activation of DNA repair mechanisms and a persistent DNA damage response (DDR) (Victorelli and Passos 2017). Nevertheless, telomere dysfunction can occur irrespective of length with telomeric DNA damage being associated with an increase in senescence markers such as p16 (Victorelli and Passos 2017; Birch et al. 2015). Indeed, in postmitotic cardiomyocytes, there is an increase in DNA damage foci associated with telomeres during ageing (Anderson et al. 2019). Telomeres interact with a telomere repeat-binding factors 1 and 2 (TRF1 and TRF2, respectively) to form t-loops. TRF1 is thought to prevent fusion of telomere ends and regulate telomere length (van Steensel et al. 1998; Celli and de Lange 2005), whilst TRF2 forestalls the DNA damage response (Karlseder et al. 2004).

The positioning of telomeres in interphase nuclei appears to vary between species, cell type and disease status (Weierich et al. 2003; Chuang et al. 2004; Arnoult et al. 2010; Gilson et al. 2013). Nevertheless, positioning is non-random and integral to genomic stability (Chuang et al. 2004). There is evidence that telomeres are closely associated with the nucleoskeleton and A-type lamins (Ottaviani et al. 2009; de Lange 1992) in addition to reports that the telomeres of acrocentric chromosomes localise to perinucleolar regions (Ramirez and Surralles 2008). Loss of TRF2 has been linked to an increased DNA damage response and senescence (van Steensel et al. 1998; Okamoto et al. 2013). TRF2 may interact with lamin A/C, which are important proteins within the INM and nuclear matrix (Wood et al. 2014). In HGPS, there is a reduction in TRF2 (Wood et al. 2014) and apparent telomere loss. Interestingly, studies using human telomerase reverse transcriptase (hTERT) to evade replicative senescence on proliferating fibroblasts and HGPS cells have shown dramatic genome reorganisation with the mislocalisation of whole chromosomes 18 in the control cells and chromosome 18 and X in HGPS cells (Bikkul et al. 2019). Differences in telomere organisation have also been demonstrated in terminally differentiated cells or quiescent cells in culture that have been contact inhibited (Nagele et al. 2001). Here, interphase nuclei exhibit close telomeric associations within quiescent, non-cycling cells compared with proliferating cells (Nagele et al. 2001). Clustering of telomeres has also been demonstrated in mouse embryonic fibroblasts, as well as partial association with centromeric clusters and promyelocytic leukaemia bodies (PML) (Molenaar et al. 2003; Weierich et al. 2003; Uhlirova et al. 2010).

Epigenetic changes also accompany telomere maintenance during ageing (Uhlirova et al. 2010). Levels of H3K9me3, H4K20me3 and HP1 protein have been shown to decrease in HGPS (Scaffidi and Misteli 2006). Treatment of embryonic mouse fibroblast with the histone deacetylase inhibitor Trichostatin A (TSA) led to the repositioning of telomeres to the nuclear interior and centromeres towards the nuclear periphery (Uhlirova et al. 2010). This has been observed in another system where telomere and centromeres were often polarised to opposite ends of the chromosome territories (Amrichova et al. 2003). This may have further ramifications as genes located in close proximity to telomeric heterochromatin are often silenced due to the "telomeric position effect" (TPE) (Baur et al. 2001; Ning et al. 2003). There is evidence of expression changes in telomeric genes during senescence (Ning et al. 2003) with increased expression of the 16q telomeric genes MGC3101 and CPNE7 in senescence and GAS11 and CDK10 in both senescent/quiescent cells (Ning et al. 2003). Thus, change in the epigenetic status of constitutive heterochromatin could lead to senescent-specific expression patterns.

Chromosomes and Chromosome Territories

In interphase nuclei, whole chromosomes occupy specific non-random locations within the nuclear space called chromosome territories (CTs), occupying similar locations between cell types and in vitro compared to ex vivo (Foster et al. 2012). In proliferating human fibroblasts, CTs are functionally compartmentalised with gene-rich chromosomes occupying a central position within the nucleus and are generally characterised as having higher levels of gene expression, open chromatin conformations and early replication timing (Croft et al. 1999; Cremer and Cremer 2001; Foster and Bridger 2005; Bridger et al. 2014). Conversely, gene-poor chromosomes are commonly associated with the nuclear periphery or nucleoli and are synonymous with heterochromatin, repression of gene expression, repressive histone modifications and late replication timing (Chiang et al. 2018; Croft et al. 1999). These functionally different compartments have been shown to be of importance during ageing. There is evidence that whole chromosome territories can occupy different nuclear locations during senescence (Bridger et al. 2000; Mehta et al. 2007). For instance, human chromosome 18 has been shown to occupy a peripheral position within proliferating fibroblast nuclei; however, upon replicative senescence, chromosome 18 was shown to be repositioned away from the nuclear periphery (Bridger et al. 2000). Therefore, CTs appear to be repositioned from a gene-density radial distribution in proliferating fibroblasts to a size-correlated radial position within senescent fibroblasts whereby small chromosomes are positioned within the nuclear interior and large chromosomes are localised at the nuclear periphery (Mehta et al. 2007). Altered nuclear positioning of whole chromosomes 13 and 18 from the nuclear periphery towards the interior has also been shown in cells with A-type lamins (Meaburn et al. 2005; Meaburn et al. 2007). Indeed, genome reorganisation is more finely observed with changes in the Topologically Associated Domain (TADs) compartment re-positioning in replicative senescent cells (Criscione et al. 2016; Sun et al. 2018).

Advances in Technologies

New technologies including CRISPR-multicolour (Ma et al. 2015) and CRISPRainbow (Ma et al. 2016) enable the study of higher-order chromatin and nuclear architecture organisation. Here, a live-cell system that utilises super-resolution microscopy is used to track genomic loci that are labelled by different coloured fluorescent-tagged dCas9-sgRNAs (Ma et al. 2016). Changes in transcriptional activity upon a specific stimulus can also be investigated by tracking the dynamics of a promoter and its interaction with cis-/trans-acting regulatory elements (Lau and Suh 2017). New advances in interrogating Hi-C data from senescent cells are permitting chromosome territory postions to be extrapolated from these data sets (Das et al. 2020).

Summary

The complexity of the ageing genome and structural organisation of the nucleus during senescence are becoming increasingly apparent, especially with advances in microscopy and global analyses such as super-resolution microscopy and chromosome conformation capture. Generally, the ageing epigenome is characterised by global hypomethylation and loss of heterochromatin; however, on the contrary, some regions of the genome are packaged into heterochromatin, e.g. SAHF. Satellite sequences may also be altered in senescence with distension of centromeres, or SADS, and shortening or dysfunction of telomeres. Characteristic structural changes to the nucleus include an increased size in senescent cells, reorganisation of LADs and sub-NADs and nuclear envelope deformities associated with mutations in lamin and lamin-associated proteins. Together, these can lead to large-scale reorganisation of the genome with repositioning of whole chromosome territories. Overall, these fundamental changes to the epigenome lead to alterations in global gene expression and genomic instability associated with ageing (Fig. 5.2).

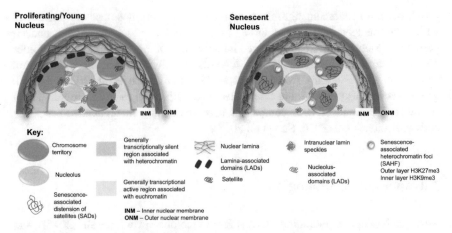

Fig. 5.2 Differences in nuclear structure organisation within proliferating/young cells and senescent cells. Senescence is associated with chromatin remodelling, loss of peripheral nuclear heterochromatin and an increase in hypomethylation. In addition, satellite DNA becomes unravelled to form SADs, and LADs and NADs are redistributed in senescent cells. Nucleoli may fuse and often have an increased size. SAHF formation is apparent in some senescence cells with a chromatin core enriched in H3K9me3 surrounded by an outer rim rich in H3K27me3. The nuclear lamina contains A- and B-type lamins and lamina-associated proteins that play a role in organising the genome. Within senescent and progeroid cells, the INM organisation is altered and often associated with nuclear envelope deformities

References

Ahringer J, Gasser SM (2018) Repressive chromatin in Caenorhabditis elegans: establishment, composition, and function. Genetics 208(2):491–511. https://doi.org/10.1534/genetics.117.300386

Aird KM, Zhang R (2013) Detection of senescence-associated heterochromatin foci (SAHF). Methods Mol Biol (Clifton, NJ) 965:185–196. https://doi.org/10.1007/978-1-62703-239-1_12

Amrichova J, Lukasova E, Kozubek S, Kozubek M (2003) Nuclear and territorial topography of chromosome telomeres in human lymphocytes. Exp Cell Res 289(1):11–26. https://doi.org/10.1016/s0014-4827(03)00208-8

Anderson R, Lagnado A, Maggiorani D, Walaszczyk A, Dookun E, Chapman J, Birch J, Salmonowicz H, Ogrodnik M, Jurk D, Proctor C, Correia-Melo C, Victorelli S, Fielder E, Berlinguer-Palmini R, Owens A, Greaves LC, Kolsky KL, Parini A, Douin-Echinard V, LeBrasseur NK, Arthur HM, Tual-Chalot S, Schafer MJ, Roos CM, Miller JD, Robertson N, Mann J, Adams PD, Tchkonia T, Kirkland JL, Mialet-Perez J, Richardson GD, Passos JF (2019) Length-independent telomere damage drives post-mitotic cardiomyocyte senescence. EMBO J 38(5):e100492. https://doi.org/10.15252/embj.2018100492

Arnoult N, Schluth-Bolard C, Letessier A, Drascovic I, Bouarich-Bourimi R, Campisi J, S-h K, Boussouar A, Ottaviani A, Magdinier F, Gilson E, Londoño-Vallejo A (2010) Replication timing of human telomeres is chromosome arm–specific, influenced by subtelomeric structures and connected to nuclear localization. PLoS Genet 6(4):e1000920. https://doi.org/10.1371/journal.pgen.1000920

Ashapkin VV, Kutueva LI, Kurchashova SY, Kireev II (2019) Are there common mechanisms between the Hutchinson-Gilford progeria syndrome and natural aging? Front Genet 10:455. https://doi.org/10.3389/fgene.2019.00455

Baillie JK, Barnett MW, Upton KR, Gerhardt DJ, Richmond TA, De Sapio F, Brennan PM, Rizzu P, Smith S, Fell M, Talbot RT, Gustincich S, Freeman TC, Mattick JS, Hume DA, Heutink P, Carninci P, Jeddeloh JA, Faulkner GJ (2011) Somatic retrotransposition alters the genetic landscape of the human brain. Nature 479(7374):534–537. https://doi.org/10.1038/nature10531

Barboro P, Repaci E, D'Arrigo C, Balbi C (2012) The role of nuclear matrix proteins binding to matrix attachment regions (Mars) in prostate cancer cell differentiation. PLoS One 7(7):e40617. https://doi.org/10.1371/journal.pone.0040617

Baur JA, Zou Y, Shay JW, Wright WE (2001) Telomere position effect in human cells. Science 292(5524):2075–2077. https://doi.org/10.1126/science.1062329

Bersaglieri C, Santoro R (2019) Genome organization in and around the nucleolus. Cell 8(6):579. https://doi.org/10.3390/cells8060579

Bikkul MU, Clements CS, Godwin LS, Goldberg MW, Kill IR, Bridger JM (2018) Farnesyltransferase inhibitor and rapamycin correct aberrant genome organisation and decrease DNA damage respectively, in Hutchinson-Gilford progeria syndrome fibroblasts. Biogerontology 19(6):579–602. https://doi.org/10.1007/s10522-018-9758-4

Bikkul MU, Faragher RGA, Worthington G, Meinke P, Kerr ARW, Sammy A, Riyahi K, Horton D, Schirmer EC, Hubank M, Kill IR, Anderson RM, Slijepcevic P, Makarov E, Bridger JM (2019) Telomere elongation through hTERT immortalization leads to chromosome repositioning in control cells and genomic instability in Hutchinson-Gilford progeria syndrome fibroblasts, expressing a novel SUN1 isoform. Genes Chromosomes Cancer 58(6):341–356. https://doi.org/10.1002/gcc.22711

Birch J, Anderson RK, Correia-Melo C, Jurk D, Hewitt G, Marques FM, Green NJ, Moisey E, Birrell MA, Belvisi MG, Black F, Taylor JJ, Fisher AJ, De Soyza A, Passos JF (2015) DNA damage response at telomeres contributes to lung aging and chronic obstructive pulmonary disease. Am J Physiol Lung Cell Mol Physiol 309(10):L1124–L1137. https://doi.org/10.1152/ajplung.00293.2015

Black JC, Van Rechem C, Whetstine JR (2012) Histone lysine methylation dynamics: establishment, regulation, and biological impact. Mol Cell 48(4):491–507. https://doi.org/10.1016/j.molcel.2012.11.006

Boeger H, Griesenbeck J, Strattan JS, Kornberg RD (2003) Nucleosomes unfold completely at a transcriptionally active promoter. Mol Cell 11(6):1587–1598

Booth LN, Brunet A (2016) The aging epigenome. Mol Cell 62(5):728–744. https://doi.org/10.1016/j.molcel.2016.05.013

Boumendil C, Hari P, Olsen KCF, Acosta JC, Bickmore WA (2019) Nuclear pore density controls heterochromatin reorganization during senescence. Genes Dev 33(3–4):144–149. https://doi.org/10.1101/gad.321117.118

Braig M, Lee S, Loddenkemper C, Rudolph C, Peters AHFM, Schlegelberger B, Stein H, Dörken B, Jenuwein T, Schmitt CA (2005) Oncogene-induced senescence as an initial barrier in lymphoma development. Nature 436(7051):660–665. https://doi.org/10.1038/nature03841

Bridger JM, Kill IR, O'Farrell M, Hutchison CJ (1993) Internal Lamin structures within G1 nuclei of human dermal fibroblasts. J Cell Sci 104(Pt 2):297–306

Bridger JM, Boyle S, Kill IR, Bickmore WA (2000) Re-modelling of nuclear architecture in quiescent and senescent human fibroblasts. Curr Biol 10(3):149–152

Bridger JM, Arican-Gotkas HD, Foster HA, Godwin LS, Harvey A, Kill IR, Knight M, Mehta IS, Ahmed MH (2014) The non-random repositioning of whole chromosomes and individual gene loci in interphase nuclei and its relevance in disease, infection, aging, and cancer. Adv Exp Med Biol 773:263–279. https://doi.org/10.1007/978-1-4899-8032-8_12

Cau P, Navarro C, Harhouri K, Roll P, Sigaudy S, Kaspi E, Perrin S, De Sandre-Giovannoli A, Levy N (2014) Nuclear matrix, nuclear envelope and premature aging syndromes in a translational research perspective. Semin Cell Dev Biol 29:125–147. https://doi.org/10.1016/j.semcdb.2014.03.021

Celli GB, de Lange T (2005) DNA processing is not required for ATM-mediated telomere damage response after TRF2 deletion. Nat Cell Biol 7(7):712–718. https://doi.org/10.1038/ncb1275

Chandra T, Narita M (2013) High-order chromatin structure and the epigenome in SAHFs. Nucleus (Austin, Tex) 4(1):23–28. https://doi.org/10.4161/nucl.23189

Chandra T, Kirschner K, Thuret JY, Pope BD, Ryba T, Newman S, Ahmed K, Samarajiwa SA, Salama R, Carroll T, Stark R, Janky R, Narita M, Xue L, Chicas A, Nunez S, Janknecht R, Hayashi-Takanaka Y, Wilson MD, Marshall A, Odom DT, Babu MM, Bazett-Jones DP, Tavare S, Edwards PA, Lowe SW, Kimura H, Gilbert DM, Narita M (2012) Independence of repressive histone marks and chromatin compaction during senescent heterochromatic layer formation. Mol Cell 47(2):203–214. https://doi.org/10.1016/j.molcel.2012.06.010

Chandra T, Ewels PA, Schoenfelder S, Furlan-Magaril M, Wingett SW, Kirschner K, Thuret J-Y, Andrews S, Fraser P, Reik W (2015) Global reorganization of the nuclear landscape in senescent cells. Cell Rep 10(4):471–483. https://doi.org/10.1016/j.celrep.2014.12.055

Chattopadhyay S, Pavithra L (2007) MARs and MARBPs: key modulators of gene regulation and disease manifestation. Subcell Biochem 41:213–230

Chiang M, Michieletto D, Brackley CA, Rattanavirotkul N, Mohammed H, Marenduzzo D, Chandra T (2018) Lamina and heterochromatin direct chromosome organisation in senescence and progeria. bioRxiv:468561. https://doi.org/10.1101/468561

Chuang TCY, Moshir S, Garini Y, Chuang AY-C, Young IT, Vermolen B, van den Doel R, Mougey V, Perrin M, Braun M, Kerr PD, Fest T, Boukamp P, Mai S (2004) The three-dimensional organization of telomeres in the nucleus of mammalian cells. BMC Biol 2:12–12. https://doi.org/10.1186/1741-7007-2-12

Coppé J-P, Desprez P-Y, Krtolica A, Campisi J (2010) The senescence-associated secretory phenotype: the dark side of tumor suppression. Annu Rev Pathol 5:99–118. https://doi.org/10.1146/annurev-pathol-121808-102144

Cremer T, Cremer C (2001) Chromosome territories, nuclear architecture and gene regulation in mammalian cells. Nat Rev Genet 2(4):292–301. https://doi.org/10.1038/35066075

Criscione SW, De Cecco M, Siranosian B, Zhang Y, Kreiling JA, Sedivy JM, Neretti N (2016) Reorganization of chromosome architecture in replicative cellular senescence. Sci Adv 2(2):e1500882. https://doi.org/10.1126/sciadv.1500882

Croft JA, Bridger JM, Boyle S, Perry P, Teague P, Bickmore WA (1999) Differences in the localization and morphology of chromosomes in the human nucleus. J Cell Biol 145(6):1119–1131. https://doi.org/10.1083/jcb.145.6.1119

Cruickshanks HA, McBryan T, Nelson DM, Vanderkraats ND, Shah PP, van Tuyn J, Singh Rai T, Brock C, Donahue G, Dunican DS, Drotar ME, Meehan RR, Edwards JR, Berger SL, Adams PD (2013) Senescent cells harbour features of the cancer epigenome. Nat Cell Biol 15(12):1495–1506. https://doi.org/10.1038/ncb2879

Das P, Shen T, McCord RP (2020) Inferring chromosome radial organization from Hi-C data. BMC Bioinf 21(1):511

De Cecco M, Criscione SW, Peckham EJ, Hillenmeyer S, Hamm EA, Manivannan J, Peterson AL, Kreiling JA, Neretti N, Sedivy JM (2013) Genomes of replicatively senescent cells undergo global epigenetic changes leading to gene silencing and activation of transposable elements. Aging Cell 12(2):247–256. https://doi.org/10.1111/acel.12047

de Koning AP, Gu W, Castoe TA, Batzer MA, Pollock DD (2011) Repetitive elements may comprise over two-thirds of the human genome. PLoS Genet 7(12):e1002384. https://doi.org/10.1371/journal.pgen.1002384

de Lange T (1992) Human telomeres are attached to the nuclear matrix. EMBO J 11(2):717–724

Dillinger S, Straub T, Németh A (2017) Nucleolus association of chromosomal domains is largely maintained in cellular senescence despite massive nuclear reorganisation. PLoS One 12(6):e0178821. https://doi.org/10.1371/journal.pone.0178821

Dupont C, Armant DR, Brenner CA (2009) Epigenetics: definition, mechanisms and clinical perspective. Semin Reprod Med 27(5):351–357. https://doi.org/10.1055/s-0029-1237423

Ehrlich M (2019) DNA hypermethylation in disease: mechanisms and clinical relevance. Epigenetics:1–23. https://doi.org/10.1080/15592294.2019.1638701

Elbarbary RA, Lucas BA, Maquat LE (2016) Retrotransposons as regulators of gene expression. Science 351(6274):aac7247. https://doi.org/10.1126/science.aac7247

Elcock LS, Bridger JM (2008) Exploring the effects of a dysfunctional nuclear matrix. Biochem Soc Transactions 36:1378–83

Foster HA, Bridger JM (2005) The genome and the nucleus: a marriage made by evolution. Genome organisation and nuclear architecture. Chromosoma 114(4):212–229. https://doi.org/10.1007/s00412-005-0016-6

Foster HA, Griffin DK, Bridger JM (2012) Interphase chromosome positioning in in vitro porcine cells and ex vivo porcine tissues. BMC Cell Biol 13:30. https://doi.org/10.1186/1471-2121-13-30

Galvani A, Courbeyrette R, Agez M, Ochsenbein F, Mann C, Thuret J-Y (2008) In vivo study of the nucleosome assembly functions of ASF1 histone chaperones in human cells. Mol Cell Biol 28(11):3672–3685. https://doi.org/10.1128/MCB.00510-07

Gensous N, Bacalini MG, Pirazzini C, Marasco E, Giuliani C, Ravaioli F, Mengozzi G, Bertarelli C, Palmas MG, Franceschi C, Garagnani P (2017) The epigenetic landscape of age-related diseases: the geroscience perspective. Biogerontology 18(4):549–559. https://doi.org/10.1007/s10522-017-9695-7

Gilbert HTJ, Swift J (2019) The consequences of ageing, progeroid syndromes and cellular senescence on mechanotransduction and the nucleus. Exp Cell Res 378(1):98–103. https://doi.org/10.1016/j.yexcr.2019.03.002

Gilson E, Giraud-Panis M-J, Pisano S, Bennaroch D, Ledu M-H, Pei B (2013) One identity or more for telomeres? Front Oncol 3(48). https://doi.org/10.3389/fonc.2013.00048

Godwin LS, Bridger JM, Foster HA (2021) Fluorescence in situ hybridization on DNA Halo preparations to reveal whole chromosome, telomeres and gene loci. Jove, in press

Goldman RD, Gruenbaum Y, Moir RD, Shumaker DK, Spann TP (2002) Nuclear lamins: building blocks of nuclear architecture. Genes Dev 16(5):533–547. https://doi.org/10.1101/gad.960502

Goldman RD, Shumaker DK, Erdos MR, Eriksson M, Goldman AE, Gordon LB, Gruenbaum Y, Khuon S, Mendez M, Varga R, Collins FS (2004) Accumulation of mutant lamin A causes progressive changes in nuclear architecture in Hutchinson-Gilford progeria syndrome. Proc Natl Acad Sci U S A 101(24):8963–8968. https://doi.org/10.1073/pnas.0402943101

Graziano S, Gonzalo S (2017) Mechanisms of oncogene-induced genomic instability. Biophys Chem 225:49–57. https://doi.org/10.1016/j.bpc.2016.11.008

Haithcock E, Dayani Y, Neufeld E, Zahand AJ, Feinstein N, Mattout A, Gruenbaum Y, Liu J (2005) Age-related changes of nuclear architecture in Caenorhabditis elegans. Proc Natl Acad Sci U S A 102(46):16690–16695. https://doi.org/10.1073/pnas.0506955102

Haugstetter AM, Loddenkemper C, Lenze D, Gröne J, Standfuß C, Petersen I, Dörken B, Schmitt CA (2010) Cellular senescence predicts treatment outcome in metastasised colorectal cancer. Br J Cancer 103:505. https://doi.org/10.1038/sj.bjc.6605784

Hayflick L (1965) The limited in vitro lifetime of human diploid cell strains. Exp Cell Res 37:614–636

Heyn H, Li N, Ferreira HJ, Moran S, Pisano DG, Gomez A, Diez J, Sanchez-Mut JV, Setien F, Carmona FJ, Puca AA, Sayols S, Pujana MA, Serra-Musach J, Iglesias-Platas I, Formiga F, Fernandez AF, Fraga MF, Heath SC, Valencia A, Gut IG, Wang J, Esteller M (2012) Distinct DNA methylomes of newborns and centenarians. Proc Natl Acad Sci U S A 109(26):10522–10527. https://doi.org/10.1073/pnas.1120658109

Horvath S (2013) DNA methylation age of human tissues and cell types. Genome Biol 14(10):R115. https://doi.org/10.1186/gb-2013-14-10-r115

Horvath S, Raj K (2018) DNA methylation-based biomarkers and the epigenetic clock theory of ageing. Nat Rev Genet 19(6):371–384. https://doi.org/10.1038/s41576-018-0004-3

Hu Z, Chen K, Xia Z, Chavez M, Pal S, Seol J-H, Chen C-C, Li W, Tyler JK (2014) Nucleosome loss leads to global transcriptional up-regulation and genomic instability during yeast aging. Genes Dev 28(4):396–408. https://doi.org/10.1101/gad.233221.113

Jeziorska DM, Murray RJS, De Gobbi M, Gaentzsch R, Garrick D, Ayyub H, Chen T, Li E, Telenius J, Lynch M, Graham B, Smith AJH, Lund JN, Hughes JR, Higgs DR, Tufarelli C (2017) DNA methylation of intragenic CpG islands depends on their transcriptional activity during differentiation and disease. Proc Natl Acad Sci 114(36):E7526. https://doi.org/10.1073/pnas.1703087114

Karlseder J, Hoke K, Mirzoeva OK, Bakkenist C, Kastan MB, Petrini JH, de Lange T (2004) The telomeric protein TRF2 binds the ATM kinase and can inhibit the ATM-dependent DNA damage response. PLoS Biol 2(8):E240. https://doi.org/10.1371/journal.pbio.0020240

Keyes WM (2013) Rearranging senescence: transposable elements become active in aging cells (comment on DOI 10.1002/bies.201300097). Bioessays 35(12):1023–1023. https://doi.org/10.1002/bies.201300157

Larson K, Yan S-J, Tsurumi A, Liu J, Zhou J, Gaur K, Guo D, Eickbush TH, Li WX (2012) Heterochromatin formation promotes longevity and represses ribosomal RNA synthesis. PLoS Genet 8(1):e1002473. https://doi.org/10.1371/journal.pgen.1002473

Lau CH, Suh Y (2017) Genome and epigenome editing in mechanistic studies of human aging and aging-related disease. Gerontology 63(2):103–117. https://doi.org/10.1159/000452972

Lazzerini Denchi E, Attwooll C, Pasini D, Helin K (2005) Deregulated E2F activity induces hyperplasia and senescence-like features in the mouse pituitary gland. Mol Cell Biol 25(7):2660–2672. https://doi.org/10.1128/mcb.25.7.2660-2672.2005

Lenain C, de Graaf CA, Pagie L, Visser NL, de Haas M, de Vries SS, Peric-Hupkes D, van Steensel B, Peeper DS (2017) Massive reshaping of genome-nuclear lamina interactions during oncogene-induced senescence. Genome Res 27(10):1634–1644. https://doi.org/10.1101/gr.225763.117

Levine ME, Lu AT, Bennett DA, Horvath S (2015) Epigenetic age of the pre-frontal cortex is associated with neuritic plaques, amyloid load, and Alzheimer's disease related cognitive functioning. Aging (Albany NY) 7(12):1198–1211. https://doi.org/10.18632/aging.100864

Lochs SJA, Kefalopoulou S, Kind J (2019) Lamina associated domains and gene regulation in development and cancer. Cell 8(3):271

Lukasova E, Kovarik A, Kozubek S (2018) Consequences of Lamin B1 and Lamin B receptor downregulation in senescence. Cell 7(2). https://doi.org/10.3390/cells7020011

Ma H, Naseri A, Reyes-Gutierrez P, Wolfe SA, Zhang S, Pederson T (2015) Multicolor CRISPR labeling of chromosomal loci in human cells. Proc Natl Acad Sci U S A 112(10):3002–3007. https://doi.org/10.1073/pnas.1420024112

Ma H, Tu L-C, Naseri A, Huisman M, Zhang S, Grunwald D, Pederson T (2016) Multiplexed labeling of genomic loci with dCas9 and engineered sgRNAs using CRISPRainbow. Nat Biotechnol 34(5):528–530. https://doi.org/10.1038/nbt.3526

Maleszewska M, Mawer JSP, Tessarz P (2016) Histone modifications in ageing and lifespan regulation. Curr Mol Biol Rep 2(1):26–35. https://doi.org/10.1007/s40610-016-0031-9

McClintock D, Ratner D, Lokuge M, Owens DM, Gordon LB, Collins FS, Djabali K (2007) The mutant form of Lamin a that causes Hutchinson-Gilford progeria is a biomarker of cellular aging in human skin. PLoS One 2(12):e1269. https://doi.org/10.1371/journal.pone.0001269

Meaburn KJ, Levy N, Toniolo D, Bridger JM (2005) Chromosome positioning is largely unaffected in lymphoblastoid cell lines containing emerin or A-type Lamin mutations. Biochem Soc Trans 33(Pt 6):1438–1440

Meaburn KJ, Cabuy E, Bonne G, Levy N, Morris GE, Novelli G, Kill IR, Bridger JM (2007) Primary laminopathy fibroblasts display altered genome organization and apoptosis. Aging Cell 6(2):139–153. https://doi.org/10.1111/j.1474-9726.2007.00270.x

Medstrand P, van de Lagemaat LN, Mager DL (2002) Retroelement distributions in the human genome: variations associated with age and proximity to genes. Genome Res 12(10):1483–1495. https://doi.org/10.1101/gr.388902

Mehta IS, Figgitt M, Clements CS, Kill IR, Bridger JM (2007) Alterations to nuclear architecture and genome behavior in senescent cells. Ann N Y Acad Sci 1100:250–263. https://doi.org/10.1196/annals.1395.027

Mehta IS, Bridger JM, Kill IR (2010) Progeria, the nucleolus and farnesyltransferase inhibitors. Biochem Soc Trans 38(Pt 1):287–291. https://doi.org/10.1042/bst0380287

Meuleman W, Peric-Hupkes D, Kind J, Beaudry J-B, Pagie L, Kellis M, Reinders M, Wessels L, van Steensel B (2013) Constitutive nuclear lamina-genome interactions are highly conserved and associated with A/T-rich sequence. Genome Res 23(2):270–280. https://doi.org/10.1101/gr.141028.112

Michaloglou C, Vredeveld LCW, Soengas MS, Denoyelle C, Kuilman T, van der Horst CMAM, Majoor DM, Shay JW, Mooi WJ, Peeper DS (2005) BRAFE600-associated senescence-like cell cycle arrest of human naevi. Nature 436(7051):720–724. https://doi.org/10.1038/nature03890

Mills RE, Bennett EA, Iskow RC, Luttig CT, Tsui C, Pittard WS, Devine SE (2006) Recently mobilized transposons in the human and chimpanzee genomes. Am J Hum Genet 78(4):671–679. https://doi.org/10.1086/501028

Miranda-Morales E, Meier K, Sandoval-Carrillo A, Salas-Pacheco J, Vázquez-Cárdenas P, Arias-Carrión O (2017) Implications of DNA methylation in Parkinson's disease. Front Mol Neurosci 10:225–225. https://doi.org/10.3389/fnmol.2017.00225

Molenaar C, Wiesmeijer K, Verwoerd NP, Khazen S, Eils R, Tanke HJ, Dirks RW (2003) Visualizing telomere dynamics in living mammalian cells using PNA probes. EMBO J 22(24):6631–6641. https://doi.org/10.1093/emboj/cdg633

Morris BJ, Willcox BJ, Donlon TA (2019) Genetic and epigenetic regulation of human aging and longevity. Biochim Biophys Acta (BBA) – Mol Basis Dis 1865(7):1718–1744. https://doi.org/10.1016/j.bbadis.2018.08.039

Munoz-Espin D, Canamero M, Maraver A, Gomez-Lopez G, Contreras J, Murillo-Cuesta S, Rodriguez-Baeza A, Varela-Nieto I, Ruberte J, Collado M, Serrano M (2013) Programmed cell senescence during mammalian embryonic development. Cell 155(5):1104–1118. https://doi.org/10.1016/j.cell.2013.10.019

Musich PR, Zou Y (2009) Genomic instability and DNA damage responses in progeria arising from defective maturation of prelamin A. Aging (Albany NY) 1(1):28–37. https://doi.org/10.18632/aging.100012

Nagai K, Matsushita T, Matsuzaki T, Takayama K, Matsumoto T, Kuroda R, Kurosaka M (2015) Depletion of SIRT6 causes cellular senescence, DNA damage, and telomere dysfunction in human chondrocytes. Osteoarthr Cartil 23(8):1412–1420. https://doi.org/10.1016/j.joca.2015.03.024

Nagele RG, Velasco AQ, Anderson WJ, McMahon DJ, Thomson Z, Fazekas J, Wind K, Lee H (2001) Telomere associations in interphase nuclei: possible role in maintenance of interphase chromosome topology. J Cell Sci 114(2):377

Navarro-Sánchez L, Águeda-Gómez B, Aparicio S, Pérez-Tur J (2018) Epigenetic study in Parkinson's disease: a pilot analysis of DNA methylation in candidate genes in brain. Cell 7(10):150. https://doi.org/10.3390/cells7100150

Nemeth A, Conesa A, Santoyo-Lopez J, Medina I, Montaner D, Peterfia B, Solovei I, Cremer T, Dopazo J, Langst G (2010) Initial genomics of the human nucleolus. PLoS Genet 6(3):e1000889. https://doi.org/10.1371/journal.pgen.1000889

Ni Z, Ebata A, Alipanahiramandi E, Lee SS (2012) Two SET domain containing genes link epigenetic changes and aging in Caenorhabditis elegans. Aging Cell 11(2):315–325. https://doi.org/10.1111/j.1474-9726.2011.00785.x

Ning Y, Xu J-f, Li Y, Chavez L, Riethman HC, Lansdorp PM, N-p W (2003) Telomere length and the expression of natural telomeric genes in human fibroblasts. Hum Mol Genet 12(11):1329–1336. https://doi.org/10.1093/hmg/ddg139

O'Donnell KA, Burns KH (2010) Mobilizing diversity: transposable element insertions in genetic variation and disease. Mob DNA 1(1):21–21. https://doi.org/10.1186/1759-8753-1-21

O'Sullivan RJ, Kubicek S, Schreiber SL, Karlseder J (2010) Reduced histone biosynthesis and chromatin changes arising from a damage signal at telomeres. Nat Struct Mol Biol 17(10):1218–1225. https://doi.org/10.1038/nsmb.1897

Ogneva ZV, Dubrovina AS, Kiselev KV (2016) Age-associated alterations in DNA methylation and expression of methyltransferase and demethylase genes in Arabidopsis thaliana. Biol Plant 60(4):628–634. https://doi.org/10.1007/s10535-016-0638-y

Okamoto K, Bartocci C, Ouzounov I, Diedrich JK, Yates JR 3rd, Denchi EL (2013) A two-step mechanism for TRF2-mediated chromosome-end protection. Nature 494(7438):502–505. https://doi.org/10.1038/nature11873

Ottaviani A, Schluth-Bolard C, Rival-Gervier S, Boussouar A, Rondier D, Foerster AM, Morere J, Bauwens S, Gazzo S, Callet-Bauchu E, Gilson E, Magdinier F (2009) Identification of a perinuclear positioning element in human subtelomeres that requires A-type lamins and CTCF. EMBO J 28(16):2428–2436. https://doi.org/10.1038/emboj.2009.201

Park J-W, Kim JJ, Bae Y-S (2018) CK2 downregulation induces senescence-associated heterochromatic foci formation through activating SUV39h1 and inactivating G9a. Biochem Biophys Res Commun 505(1):67–73. https://doi.org/10.1016/j.bbrc.2018.09.099

Peric-Hupkes D, Meuleman W, Pagie L, Bruggeman SWM, Solovei I, Brugman W, Gräf S, Flicek P, Kerkhoven RM, van Lohuizen M, Reinders M, Wessels L, van Steensel B (2010) Molecular maps of the reorganization of genome-nuclear lamina interactions during differentiation. Mol Cell 38(4):603–613. https://doi.org/10.1016/j.molcel.2010.03.016

Ramirez MJ, Surralles J (2008) Laser confocal microscopy analysis of human interphase nuclei by three-dimensional FISH reveals dynamic perinucleolar clustering of telomeres. Cytogenet Genome Res 122(3–4):237–242. https://doi.org/10.1159/000167809

Romero-Bueno R, Ruiz DP, Artal-Sanz M, Askjaer P, Dobrzynska A (2019) Nuclear Organization in Stress and Aging. Cell 8(7). https://doi.org/10.3390/cells8070664

Sadaie M, Salama R, Carroll T, Tomimatsu K, Chandra T, Young AR, Narita M, Perez-Mancera PA, Bennett DC, Chong H, Kimura H, Narita M (2013) Redistribution of the Lamin B1 genomic binding profile affects rearrangement of heterochromatic domains and SAHF formation during senescence. Genes Dev 27(16):1800–1808. https://doi.org/10.1101/gad.217281.113

Scaffidi P, Misteli T (2006) Lamin A-dependent nuclear defects in human aging. Science 312(5776):1059–1063. https://doi.org/10.1126/science.1127168

Sen P, Dang W, Donahue G, Dai J, Dorsey J, Cao X, Liu W, Cao K, Perry R, Lee JY, Wasko BM, Carr DT, He C, Robison B, Wagner J, Gregory BD, Kaeberlein M, Kennedy BK, Boeke JD, Berger SL (2015) H3K36 methylation promotes longevity by enhancing transcriptional fidelity. Genes Dev 29(13):1362–1376. https://doi.org/10.1101/gad.263707.115

Shah PP, Donahue G, Otte GL, Capell BC, Nelson DM, Cao K, Aggarwala V, Cruickshanks HA, Rai TS, McBryan T, Gregory BD, Adams PD, Berger SL (2013) Lamin B1 depletion in senescent cells triggers large-scale changes in gene expression and the chromatin landscape. Genes Dev 27(16):1787–1799. https://doi.org/10.1101/gad.223834.113

Shumaker DK, Dechat T, Kohlmaier A, Adam SA, Bozovsky MR, Erdos MR, Eriksson M, Goldman AE, Khuon S, Collins FS, Jenuwein T, Goldman RD (2006) Mutant nuclear lamin A leads to progressive alterations of epigenetic control in premature aging. Proc Natl Acad Sci U S A 103(23):8703–8708. https://doi.org/10.1073/pnas.0602569103

Sidler C, Kovalchuk O, Kovalchuk I (2017) Epigenetic regulation of cellular senescence and aging. Front Genet 8:138–138. https://doi.org/10.3389/fgene.2017.00138

Song S, Johnson FB (2018) Epigenetic mechanisms impacting aging: a focus on histone levels and telomeres. Genes (Basel) 9(4):201. https://doi.org/10.3390/genes9040201

Stephens AD, Liu PZ, Banigan EJ, Almassalha LM, Backman V, Adam SA, Goldman RD, Marko JF (2018) Chromatin histone modifications and rigidity affect nuclear morphology independent of lamins. Mol Biol Cell 29(2):220–233. https://doi.org/10.1091/mbc.E17-06-0410

Storer M, Mas A, Robert-Moreno A, Pecoraro M, Ortells MC, Di Giacomo V, Yosef R, Pilpel N, Krizhanovsky V, Sharpe J, Keyes WM (2013) Senescence is a developmental mechanism that contributes to embryonic growth and patterning. Cell 155(5):1119–1130. https://doi.org/10.1016/j.cell.2013.10.041

Sturm A, Ivics Z, Vellai T (2015) The mechanism of ageing: primary role of transposable elements in genome disintegration. Cell Mol Life Sci 72(10):1839–1847. https://doi.org/10.1007/s00018-015-1896-0

Sun L, Yu R, Dang W (2018) Chromatin architectural changes during cellular senescence and aging. Genes (Basel) 9(4):pii: E211. https://doi.org/10.3390/genes9040211

Swanson EC, Manning B, Zhang H, Lawrence JB (2013) Higher-order unfolding of satellite heterochromatin is a consistent and early event in cell senescence. J Cell Biol 203(6):929–942. https://doi.org/10.1083/jcb.201306073

Tanae K, Horiuchi T, Matsuo Y, Katayama S, Kawamukai M (2012) Histone chaperone Asf1 plays an essential role in maintaining genomic stability in fission yeast. PLoS One 7(1):e30472. https://doi.org/10.1371/journal.pone.0030472

Tasselli L, Xi Y, Zheng W, Tennen RI, Odrowaz Z, Simeoni F, Li W, Chua KF (2016) SIRT6 deacetylates H3K18ac at pericentric chromatin to prevent mitotic errors and cellular senescence. Nat Struct Mol Biol 23(5):434–440. https://doi.org/10.1038/nsmb.3202

Tiwari V, Wilson DM 3rd (2019) DNA damage and associated DNA repair defects in disease and premature aging. Am J Hum Genet 105(2):237–257. https://doi.org/10.1016/j.ajhg.2019.06.005

Toiber D, Erdel F, Bouazoune K, Silberman DM, Zhong L, Mulligan P, Sebastian C, Cosentino C, Martinez-Pastor B, Giacosa S, D'Urso A, Naar AM, Kingston R, Rippe K, Mostoslavsky R (2013) SIRT6 recruits SNF2H to DNA break sites, preventing genomic instability through chromatin remodeling. Mol Cell 51(4):454–468. https://doi.org/10.1016/j.molcel.2013.06.018

Trojer P, Reinberg D (2007) Facultative heterochromatin: is there a distinctive molecular signature? Mol Cell 28(1):1–13. https://doi.org/10.1016/j.molcel.2007.09.011

Tsurumi A, Li WX (2012) Global heterochromatin loss: a unifying theory of aging? Epigenetics 7(7):680–688. https://doi.org/10.4161/epi.20540

Uhlirova R, Horakova AH, Galiova G, Legartova S, Matula P, Fojtova M, Varecha M, Amrichova J, Vondracek J, Kozubek S, Bartova E (2010) SUV39h- and A-type lamin-dependent telomere nuclear rearrangement. J Cell Biochem 109(5):915–926. https://doi.org/10.1002/jcb.22466

van Koningsbruggen S, Gierliński M, Schofield P, Martin D, Barton GJ, Ariyurek Y, Dunnen JTD, Lamond AI (2010) High-resolution whole-genome sequencing reveals that specific chromatin domains from most human chromosomes associate with nucleoli. Mol Biol Cell 21(21):3735–3748. https://doi.org/10.1091/mbc.e10-06-0508

van Steensel B, Belmont AS (2017) Lamina-associated domains: links with chromosome architecture, heterochromatin, and gene repression. Cell 169(5):780–791. https://doi.org/10.1016/j.cell.2017.04.022

van Steensel B, Smogorzewska A, de Lange T (1998) TRF2 protects human telomeres from end-to-end fusions. Cell 92(3):401–413. https://doi.org/10.1016/s0092-8674(00)80932-0

Vanrobays E (2017) Heterochromatin positioning and nuclear architecture. Annu Plant Rev 46:33. https://doi.org/10.1002/9781119312994.apr0502

Victorelli S, Passos JF (2017) Telomeres and cell senescence – size matters not. EBioMedicine 21:14–20. https://doi.org/10.1016/j.ebiom.2017.03.027

Villeponteau B (1997) The heterochromatin loss model of aging. Exp Gerontol 32(4–5):383–394

Wang J, Jia ST, Jia S (2016) New insights into the regulation of heterochromatin. Trends Genet: TIG 32(5):284–294. https://doi.org/10.1016/j.tig.2016.02.005

Wei JW, Huang K, Yang C, Kang CS (2017) Non-coding RNAs as regulators in epigenetics (review). Oncol Rep 37(1):3–9. https://doi.org/10.3892/or.2016.5236

Weierich C, Brero A, Stein S, von Hase J, Cremer C, Cremer T, Solovei I (2003) Three-dimensional arrangements of centromeres and telomeres in nuclei of human and murine lymphocytes. Chromosome Res 11(5):485–502

Wilson RHC, Coverley D (2017) Transformation-induced changes in the DNA-nuclear matrix interface, revealed by high-throughput analysis of DNA halos. Sci Rep 7(1):6475. https://doi.org/10.1038/s41598-017-06459-7

Wood AM, Rendtlew Danielsen JM, Lucas CA, Rice EL, Scalzo D, Shimi T, Goldman RD, Smith ED, Le Beau MM, Kosak ST (2014) TRF2 and lamin A/C interact to facilitate the functional organization of chromosome ends. Nat Commun 5:5467–5467. https://doi.org/10.1038/ncomms6467

Xie W, Baylin SB, Easwaran H (2019) DNA methylation in senescence, aging and cancer. Oncoscience 6(1–2):291–293. https://doi.org/10.18632/oncoscience.476

Xu J, Liu Y (2019) A guide to visualizing the spatial epigenome with super-resolution microscopy. FEBS J 0 (0). https://doi.org/10.1111/febs.14938

Zhang R, Chen W, Adams PD (2007) Molecular dissection of formation of senescence-associated heterochromatin foci. Mol Cell Biol 27(6):2343. https://doi.org/10.1128/MCB.02019-06

Zhang W, Li J, Suzuki K, Qu J, Wang P, Zhou J, Liu X, Ren R, Xu X, Ocampo A, Yuan T, Yang J, Li Y, Shi L, Guan D, Pan H, Duan S, Ding Z, Li M, Yi F, Bai R, Wang Y, Chen C, Yang F, Li X, Wang Z, Aizawa E, Goebl A, Soligalla RD, Reddy P, Esteban CR, Tang F, Liu G-H, Belmonte JCI (2015) Aging stem cells. A Werner syndrome stem cell model unveils heterochromatin alterations as a driver of human aging. Science 348(6239):1160–1163. https://doi.org/10.1126/science.aaa1356

Chapter 6
Unclassified Chromosome Abnormalities and Genome Behavior in Interphase

Christine J. Ye, Sarah Regan, Guo Liu, Batoul Abdallah, Steve Horne, and Henry H. Heng

Abstract The discovery and characterization of abnormal chromosomes have been an important tradition for cytogenetics. In the past 70 years, extensive efforts have been made to illustrate the molecular mechanisms of various chromosomal abnormalities and to apply them for clinical diagnosis and monitoring treatment responses. As a result, clinical cytogenetic analyses represent an essential component of laboratory medicine. However, efforts in both basic research and clinical implications have been focused on recurrent or clonal types of abnormalities, and the majority of non-clonal chromosome/nuclear aberrations remain unclassified and lack their deserved attention. In recent years, these stochastic genome-level alterations have become an important topic due to the emergence of the genome theory, in which chromosomal/nuclear variations play the ultimately important role both in somatic and organismal evolution. In this chapter, following a brief review of these studies on unclassified chromosomal/nuclear abnormalities, both the rationale and significance of studying these structures will be presented. Specifically, the dynamic relationship between normal and "abnormal" chromosomal structures, and among diverse types of "abnormal variations," will be discussed through the lens of genome-mediated somatic evolution. This discussion will not only enforce the importance of new genomic concepts, such as system inheritance, fuzzy inheritance, and emergent cellular behavior based on interaction among lower-level agents, but can also shine light on many current puzzling issues, such as missing

C. J. Ye (✉)
The Division of Hematology/Oncology, Department of Internal Medicine, University of Michigan, Ann Arbor, MI, USA
e-mail: jchrisye@med.umich.edu

S. Regan · G. Liu · B. Abdallah · S. Horne
Center for Molecular Medicine and Genomics, Wayne State University School of Medicine, Detroit, MI, USA

H. H. Heng
Center for Molecular Medicine and Genomics, Wayne State University School of Medicine, Detroit, MI, USA

Department of Pathology, Wayne State University School of Medicine, Detroit, MI, USA
e-mail: hheng@med.wayne.edu

© Springer Nature Switzerland AG 2020
I. Iourov et al. (eds.), *Human Interphase Chromosomes*,
https://doi.org/10.1007/978-3-030-62532-0_6

heritability and the challenge of clinical prediction based on gene mutation profiles. Together, genome-based genomic information will play an important role in future cytogenetics and cytogenomics.

Historical Perspective

Following the establishment of the correct number of human chromosomes (Tjio and Levan 1956), abnormal chromosomes were soon linked to diseases such as Down syndrome and *chronic myelocytic leukemia* or CML (Lejeune et al. 1959; Nowell and Hungerford 1960). In particular, with the introduction of various chromosomal banging methods to identify individual chromosomes (Caspersson et al. 1970), medical cytogenetics entered a new era marked by the successful identification of many known types of chromosomal abnormalities (both structural and numerical) and their linkage with an array of human diseases. Such chromosome identification capability was further strengthened due to the development of FISH technology (Langer et al. 1981; Lichter et al. 1990; Heng et al. 1991, 1992, 1997), especially once SKY (spectral karyotyping) and multiple color FISH became popular, as these techniques can rapidly and precisely identify individual chromosomes/ chromosomal regions both for mitotic and meiotic chromosomes (Speicher et al. 1996; Schröck et al. 1996; Heng et al. 2003; Ye et al. 2006). In recent years, different cytogenomic methods have also been applied to chromosomal analyses including various array and sequencing platforms (Dong et al. 2018).

Despite these technical advances, however, most of these identified chromosomal abnormalities fall in the category of recurrent or clonal types (clonal chromosome aberrations or CCAs) as they are commonly shared within patient populations. Furthermore, it is relatively easy to identify these signatures by classical cytogenetic/cytogenomic methods. According to clinical cytogenetic guidelines, "current cytogenetics defines CCAs as a given chromosome aberration which can be detected at least twice within 20 to 40 randomly examined mitotic figures. Based on this definition, the frequency of CCA needs to be higher than 5–10% in an examined cell population. In literature, however, when a CCA is reported, researchers often refer to aberrations with frequencies that are over 30%." (Heng et al. 2006a, b, 2016a).

Obviously, a large amount of "non-clonal chromosome aberrations" or NCCAs are not reported in the literature. Even though most NCCAs have a frequency of less than 10% among examined mitotic figures, the total number of them in their diverse types is enormous, given the fact that NCCAs can be detected from any individual, regardless of whether or not they are a patient. Unfortunately, however, these overwhelmingly numerous NCCAs were considered as insignificant "noise" and were largely ignored in the name of pattern identification (Mitelman 2000; Heng et al. 2006a, 2016a, b; Ye et al. 2018a).

Not surprisingly, at different fronts of genomic research, so-called genomic noise is overwhelming as well, as reflected by CNV and gene mutation profiles in patients, as well as in normal individuals (Iafrate et al. 2004; Heng 2007a, 2015, 2017a, 2019; Liehr 2016). In fact, these unexpected findings have started to challenge gene mutation theory (Heng et al. 2011a, b; Heng 2009). To illustrate this point, in this chapter, we will mainly use cytogenetic examples.

Our interests in NCCAs, including the initial descriptions of various abnormal chromosomes and nuclei, started in the early 1980s. With the discovery of free chromatin, sister unit fibers, partially or uncompleted-packing-mitotic figures or UPMs (later termed as Defective Mitotic Figures or DMFs), various nuclear fragments, and massively newly rejoined chromosomes (Heng and Chen 1985, 1988a), it was confirmed that these structures are real (rather than non-chromatin artifacts) (Heng and Shi 1997). Even though they were initially linked to drug treatments, these were clearly chromosome-related structures, which represented opportunities to study the high-order structure of the chromosome, and could be useful for monitoring different stages of the cell cycle.

Several research projects have promoted the realization of their importance, including the development of high-resolution fiber FISH and the characterization of genome chaos during cancer evolution (Heng et al. 1992, 1997). For more details, please see Heng and Shi (1997), Heng et al. (2013a, b), and Heng (2015, 2019). A number of representative examples are listed in Table 6.1.

It should be pointed out that, historically, it was highly significant when researchers could identify the linkages of these common and signature chromosomal abnormalities to various diseases, which supported the gene mutation theory of cancer and human diseases. Prior to the acceptance of the genetic basis of cancer, for example, the highly diverse chromosomal changes detected from cancer were used as evidence against the idea that cancer is caused by genetic aberrations. The identification of a specific translocation from CML and the subsequent cloning of the Bcr/Abl fusion gene have played highly significant roles in the acceptance of the gene mutation theory of cancer (Rowley 2013). Now, based on how challenging it has proven to be to identify commonly shared genetic aberrations for most cancer cases, coupled with the new realization that the majority of nonrecurrent genomic variants are of importance for somatic evolutionary potential, the new era of studying NCCAs is arriving. This transition represents an era in which it is necessary to deal with bio-complexity and uncertainty (Horne et al. 2013).

In the case of cancer research and, in particular, when studying the process of genome chaos, increased nuclear abnormalities are also linked to different types of chromosomal abnormalities and, ultimately, to CIN-mediated cancer evolution (Sheltzer et al. 2011; Siegel and Amon 2012; Zhu et al. 2012; Heng et al. 2013a, b; Heng 2015). Many interesting phenomena, including micronuclei clusters, giant nuclei, rapid nuclear fusion/fission/budding/bursting, and entosis, are now under increased investigation, leading to the realization that these abnormal nuclei can also change the chromosomal coding. In other words, genome reorganization can unify different types of chromosomal/nuclear variations under the evolutionary mechanism of genome-based selection (Heng 2015, 2019; Ye et al. 2018a, b, 2019a, b).

Table 6.1 Examples of various NCCAs reported in literature

Experiments and new concepts	Key findings	Main conclusions	Comments	References
Using drug treatment to induce elongated chromosomes in frog and human blood culture	Elevated frequencies of free chromatin, unit fibers, and DMFs were observed	They are chromatin materials rather than non-DNA contaminations. Both unit fibers and DMFs potentially represent various stages of the process of high-order structural formation. These chromosomal aberrations can be induced by drug treatment, especially within the G2 phase of the cell cycle	Despite a few publications, it failed to generate follow-up studies from others due to the reasons that the mechanism of their generation is not clear, and there is no guideline to score these structures	Heng and Chen (1985), Heng et al. (1988a, 1992, 2013a, b) and Heng (2015, 2019)
		High-resolution fiber FISH was initially developed using free chromatin and elongated chromosomes		
Using topo II inhibitors and other reagents to induce chromosomal de-condensation or DMFs	Elevated frequencies of DMFs, massive chromosome fragments, elongated chromosomes, and newly formed joined chromosomes	Various chromosomal aberrations can be induced from various cell lines	Observed mitotic cell death, genome chaos during the 1980s. But these data were held until 2004, waiting for additional mechanistic studies	Heng et al. (1988b), Haaf and Schmid (1989), Smith et al. (2001) and Heng (unpublished observations)
		Both the compromise of the G2-M checkpoint and interference with condensation is required to induce DMFs (unpublished data)		
Examining the baseline and inducibility of free chromatin and DMFs using normal individuals' blood culture	Free chromatin, C-Frag, aneuploidy, and translocations can be detected from hundreds of normal individuals with variable frequencies	Various aberrations can be observed from normal individuals, albeit at much lower frequencies	There likely is a base level of NCCAs for normal individuals	Heng et al. (2004a)

(continued)

Table 6.1 (continued)

Experiments and new concepts	Key findings	Main conclusions	Comments	References
Watching karyotype evolution in action using in vitro immortalization model	Massive chromosomal aberrations, including karyotype chaos, were observed during the punctuated phase of cancer macroevolution	In the punctuated discontinuous phase of genome evolution, there is no traceable clonal expansion between cellular generations, and the frequencies of NCCAs reach their peak	CCAs are often observed from the stepwise micro-evolutionary phase, while the peak of NCCAs is mapped into the macro-evolutionary phase	Heng et al. (2006a, b, c, (2011a, b) and Heng (2015, 2019)
Comparing frequencies of NCCAs from cell lines with different degrees of CIN; compare the baseline of NCCAs to induced NCCAs; compare the transcriptome profile of cell populations with different degrees of NCCAs; examine drug resistance from cell lines with variable degree of NCCAs	The frequencies of NCCAs are linked to the degree of CIN, transcriptome dynamics, cancer evolutionary potential, and drug resistance	NCCAs can be used as an index of CIN and evolutionary potential	NCCAs are not insignificant noise but valuable chromosomal variants	Stevens et al. (2013, 2014) and Heng et al. (2011a, b, 2013a, b)
Linking various chromosomal and nuclear abnormalities to cancer and other types of diseases	Linking aneuploidy to metastasis; describing entosis; giant nuclei in cancer; mosaicism in diseases	There are many diverse types of NCCAs; NCCAs are associated with an array of diseases	Most of the different types of abnormalities are linked by CIN	Ye et al. (2019a, b), Bloomfield and Duesberg (2016), Zhang et al. (2014), Iourov et al. (2008, 2010, 2019) and Horne et al. (2015)

(continued)

Table 6.1 (continued)

Experiments and new concepts	Key findings	Main conclusions	Comments	References
The establishment of the concepts of system inheritance and fuzzy inheritance	Cellular inheritance can be classified into gene-defined "parts inheritance" and genome-defined "system inheritance"	System inheritance represents a new type of coding which determines the gene interaction relationship. The order of genes and other DNA sequences within a chromosome and among different chromosomes provides the physical platform for gene interaction to work. The main function of sexual reproduction can maintain the chromosomal coding for a given species	System inheritance explains why chromosomal variations are important, and fuzzy inheritance explains why there are so many different types of the chromosomal variants	Heng (2009, 2015), Heng et al. (2009, 2011b, 2016a, b) and Ye et al. (2019a, b)
	By and large, genomic information is fuzzy rather than precise. This fuzziness is the genomic basis for heterogeneity	Fuzzy inheritance can be observed from multiple levels (e.g., gene and epigenetic levels) of bio-informational organization	System inheritance and fuzzy inheritance explain why it is challenging to understand missing heritability based on a gene-centric view	Heng et al. (2001, 2004b)
	There is a high level of dynamics in the chromatin loop domain during the normal cell cycle			

(continued)

Table 6.1 (continued)

Experiments and new concepts	Key findings	Main conclusions	Comments	References
Link stress and stress responses to cell death and the induced emergence of outliers following genome reorganization	While induced cell death can eliminate a large portion of cells, the induced surviving cells with new genomes can escape death and become dominant	Highly diverse genome alterations generated from different molecular mechanisms, including aneuploidy, micronuclear clusters, entosis, and chaotic genomes, share the same fact: Their genome systems have altered due to the changing of the chromosomal coding	That is the reason why it is essential to study the informational and evolutionary meaning of chromosomal variations, rather than the molecular mechanisms that lead to them, as there are so many ways to achieve new systems by altering the chromosomes	Stevens et al. (2011), Heng et al. (2011a, b), Ye et al. (2018a, b, 2019a, b), Heng et al. (2016b, 2019) and Horne et al. (2014)
Search for the evolutionary and informational mechanism of the highly diverse nuclear and chromosomal variations	The highly diverse abnormal nuclei and chromosomes can contribute to the formation of new genomes: a key strategy of survival			
Genome theory aims to unify multiple levels of genomic and non-genomic variants in both somatic and organismal evolution	Nearly all genomic variants are potentially useful for cellular adaptation, but as a trade-off, they can lead to diseases conditions	The genome is the basic unit for macroevolution	NCCAs (at the genome level) and other stochastic genomic and non-genomic alterations serve as evolutionary potential	Heng (2009, 2015, 2017a, b, 2019)

Examples of Unclassified Chromosome/Nuclear Abnormalities

As a freshly graduated student, one of us (HH) was very surprised and excited upon initially observing high frequencies of unknown chromosome/chromatin abnormalities and later realized that these high frequencies are observed even on chromosomal slides prepared from normal individuals without any special treatment. At that time, however, the majority of cytogeneticists dismissed these structures, and many considered them simply as contaminations or artifacts of slide-making. It was difficult to even publish these observations in mainstream cytogenetics journals.

A few years later, some of these elongated chromatin structures and chromosomes were used for the development of high-resolution fiber FISH (Heng et al. 1992, 1997). Despite this success, the biological meaning of these structures has been continuously ignored.

The third wave of studying these variants was triggered by the linkage of NCCAs and genome instability using various in vitro and in vivo cancer models, especially once the frequencies of NCCAs were linked to cancer evolutionary potentials (Heng et al. 2004a, 2006a, b, c). With the introduction of chromosomal coding and system inheritance, all of a sudden, it made the prefect sense to us why NCCAs are important and are detectable from normal and disease tissues but at different frequencies, and why there is a relationship among stress, cellular adaptation, system survival, and disease conditions. We have thus published accumulated data over the course of nearly three decades (Heng et al. 2004a, 2008, 2011a, b, 2013a, b; Heng 2019; Stevens et al. 2007, 2011, 2013). Furthermore, with the appreciation of fuzzy inheritance and emergent properties, more attention has been paid to the characterization and classification of different types of chromosomal/nuclear variants (Heng 2019; Ye et al. 2019a, b; Heng et al. 2019). Some examples of unclassified chromosome/nuclear abnormalities are listed below.

Free Chromatin

Free chromatin refers to those released chromatin materials detected from conventional cytogenetic preparation. They often display a spindle- or ropelike shape, and there is no apparent nuclear envelope. The generation of free chromatin can be achieved by various drug treatment and manipulating release conditions. For example, using a special high-PH buffer, an extremely long linear structure can be released (Heng et al. 1992; Heng and Tsui 1994; Heng 2000). Despite that elevated frequencies of free chromatin can be observed in some pathological conditions, even under routine slide-making conditions, the biological significance is still unclear. Potential causes might include the instability of the nuclear envelope and cell cycle checkpoints (Fig. 6.1).

Defective Mitotic Figures or DMFs

DMFs refer to partially condensed mitotic figures in which condensed chromosomes or chromosomal regions and uncondensed chromatin fibers coexist. There are three types of DMFs according to their morphological features, and the common

→

Fig. 6.1 (continued) comparison between interphase nuclei and various free chromatin generated from protocols releasing free chromatin (Heng et al. 1992). Interphase nuclei (**b** and **c**) and free chromatin (**d–i**) were prepared from a human-hamster hybrid cell line 4AF/106/KO15, which contains an altered human chromosome 7. (**b, d, f** and **h**) FISH detection results. The yellow signals represent a human chromosome (the FISH probe used is total human DNA). (**c, e, g** and **i**) Corresponding DAPI staining. From **d** to **h**, there is an increased degree of stretching. (Reused from Heng et al. 2013a)

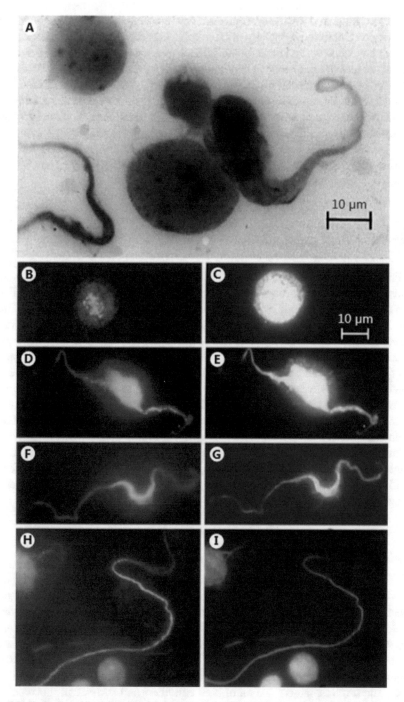

Fig. 6.1 Examples of free chromatin. (**a**) An example of the typical morphology of free chromatin (spindle and rope shapes) and three interphase nuclei detected from routine chromosome preparations without any treatment (reverse DAPI staining image). (**b**–**i**) FISH signals and morphological

feature is the mixed degree of condensation. Elevated DMFs can be obtained by using topo II inhibitor, especially in cells with a G2-M checkpoint deficiency (Heng et al. unpublished observation). Using DMF as a case study, it was realized that even the same types of chromosomal abnormalities can be linked to different errors from different phases of the cell cycle. For example, DMFs can be generated from interfering with different stages of the cell cycle, such as directly interfering with condensation in the G2 phase or indirectly interfering with DNA replication in S phase (Heng and Chen 1985; Heng et al. 1988a; Haaf and Schmid 1989; Smith et al. 2001). Even without drug treatment, the baseline of DMFs is elevated for many cancer patients, as well as in other illness conditions such as GWI and CFS (Liu et al. 2018; Heng et al. unpublished data) (Fig. 6.2).

Chromosome Fragmentations or C-Frag

C-Frags refer to the phenomenon of fragmented chromosome or nuclei. Often, different proportions of chromosomal fragments and chromosomes coexist. C-Frags represent a form of mitotic cell death (Heng et al. 2004a; Stevens et al. 2007). There are different subtypes of C-Frags based on the time fragmentation occurs (in an earlier or later stage of metaphase) and/or the degree of fragmentation (the proportion of chromosome vs. fragments). Further studies are needed to investigate if interphase nuclei can be fragmented as well. Importantly, various types of stresses (genomic and environmental alike) have been linked to the induction of C-Frags (Stevens et al. 2011; Stevens and Heng 2013), revealing the general link between various molecular pathways or mechanisms to the same end product, mitotic death. Such a connection is of importance for unifying highly diverse molecular mechanism and diverse chromosomal variations. Studies of C-Frags also help us to understand the mechanism of genome chaos (Heng et al. 2006c; Liu et al. 2014; Heng 2015, 2019). Furthermore, nuclear fragmentations are also observed (Ye et al. unpublished observations) (Fig. 6.3).

Unit Fibers

Unit fibers describe various treatment-generated (chromosomal isolation or drug treatment to interfere with condensation) substructures of metaphase chromosomes (Bak et al. 1979; Heng et al. 1988b). Unit fibers display a constant diameter of approximately 0.4 um, which have been observed from cells of different species, including frog and human. The detection of unit fibers strongly suggested that there might be an intermediate structure between metaphase and interphase chromatin fiber. The further characterization of both unit fibers and DMFs will illustrate how the last step of chromosome packaging is achieved (Heng et al. 2013a, b; Heng 2019).

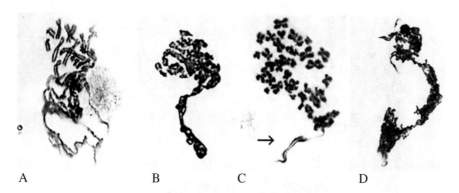

A   B   C   D

Fig. 6.2 Examples of DMFs detected from Gulf War illness patients. (**a–c**), Type 1 DMFs with the typical polarizing shape, in which the condensed chromosomes group at one end, and the uncondensed chromatin extends out in the opposite direction (Giemsa staining). In (**c**), an arrow indicates a less condensed chromosome. (**d**) Type 2 DMF with more a random distribution of de-condensed chromosomes. In this image, there is a mixture of DMFs and sticky chromosomes. (Reused from Liu et al. 2018)

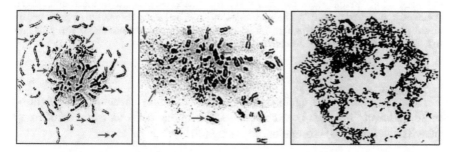

Fig. 6.3 Morphological features of chromosome fragmentation. Chromosomes undergoing fragmentation display many breaks and often seem frayed. Giemsa staining shows that chromosome fragmentation is a progressive process, with early stages showing few fragmented chromosomes (left, chromosome fragmentation (red arrows); intact chromosomes (blue arrows)), mid stage with approximately half of the chromosomes fragmented (middle), and late stage with nearly all chromosomes except for one at the top showing degradation (right). (Reused from Stevens et al. 2007)

Sticky Chromosomes

Sticky chromosomes have traditionally been described in plant chromosome research, and less attention has been paid to these structures in human chromosome studies. Sticky chromosomes can be induced by various drugs, and they are frequently observed from studies of plant hybrids. Sticky chromosomes are often observed from samples displaying high frequencies of DMFs (Heng et al. 2013b). Recently, sticky chromosomes were also detected from GWI patients (Liu et al. 2018). Sticky chromosomes can be linked to aneuploidy and translocation as well.

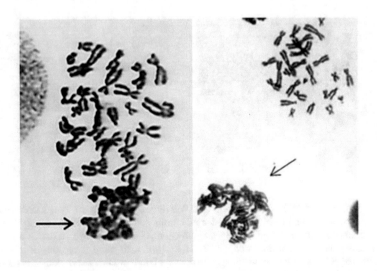

Fig. 6.4 Images of sticky chromosomes. Left: A portion of the mitotic figure displays sticky chromosomes, where multiple sticky chromosomes form a cluster (as indicated by the arrows). Right: A comparison between nonsticky chromosomes (top right) and sticky chromosomes (indicated by an arrow): This image is different from left image, as the sticky chromosome cluster likely belongs to a different mitotic figure. (Reused from Liu et al. 2018)

We also found that cells displaying high levels of sticky chromosomes might be more frequently involved in exchanging DNA among cells, an example of fuzzy inheritance. For more information, see Heng (2019) (Fig. 6.4).

Micronuclei Clusters

Unlike classical micronuclei (the small nuclei that result from chromosomes or chromosomal fragments getting separated from the daughter nucleus during cell division), the term micronuclear cluster refers to a group of various sizes of nuclei, often burst dividing from a single cell (Heng et al. 2013a, b; Heng 2019; Ye et al. 2019a). Micronuclei clusters can also be derived from giant nuclei which contain hundreds of chromosomes (Heng et al. 2013a, b, 2016a, b; Liu et al. 2014; Zhang et al. 2014; Chen et al. 2018). In a recent case study of the relationship between micronuclei and genome chaos, a general model was proposed that illustrates the mechanism of how micronuclei can promote the formation of new genome systems by reorganizing the chromosomal coding (Ye et al. 2019b) (Fig. 6.5).

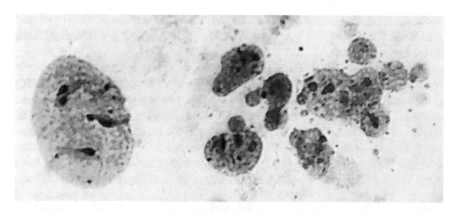

Fig. 6.5 Morphological comparison between normal interphase nucleus and micronuclei cluster. A normal nucleus is displayed at the left corner. A micronuclei cluster is located at middle to right. There are more than ten individual nuclei with different sizes (micronuclei were stained by Giemsa)

Fusion/Fission/Budding/Bursting/Entosis

Nuclei can exhibit many bizarre ways of dividing or rejoining, including cell-to-cell fusion, fission, budding, bursting, and entosis (cannibalism or emperipolesis) (Erenpreisa et al. 2005; Walen 2005; Heng 2013). On the surface, there are many differences (in regard to both morphology and mechanisms) among these many different types. Fundamentally, however, they all share the key features of altering the system inheritance or chromosomal coding and a high degree of uncertainty. Evolutionarily speaking, they all represent a stress response for cellular adaptation or survival. Despite the massive cell death involved, some outliers will have the chance to become the dominating population or serve as essential transitional populations for a new stable population to be possible. For example, entosis is a way of changing the genome through polyploidy, and polyploidy is linked to aneuploidy, translocations, and genome chaos; fusion/fission cycles are associated with genome chaos and can produce cells with altered genomes.

Chaotic Genome

This category includes many drastically altered chromosomes and nuclei (Heng et al. 2004a, 2008, 2013a, b; Liu et al. 2014; Heng 2015, 2019). For example, in addition to giant nuclei, an entire genome can form one single giant chromosome. There are chromatid rings and many other forms of alterations, most of which have yet to be named. In general, almost any form of abnormality can be detected.

It should be pointed out that chaotic genomes were initially described by cytogenetic analyses and later confirmed by sequencing. Furthermore, chromothripsis belongs to one subtype of genome chaos (Heng 2007c; Liu et al. 2011; Stephens et al. 2011; Heng et al. 2006a, b, c, 2008, 2011a; Setlur and Lee 2012; Righolt and Mai 2012; Forment et al. 2012; Crasta et al. 2012; Baca et al. 2013; Horne and Heng 2014; Liu et al. 2014).

The main reason that detections of chromothripsis have been more frequently reported than other types of genome chaos by current sequencing analysis is that these locally limited alterations can be favored by evolutionary selection and are easily detectable in clonal populations (Liu 2011; Heng et al. 2013a, b; Liu et al. 2014; Heng 2015, 2017a, b, 2019). In fact, due to the limitations of DNA sequencing (which is unable to detect cell subpopulations below 10–15%), only clonal chaotic genomes can be detected (single-cell sequencing can solve this problem, but a large number of cells are needed). In contrast, cytogenetic method is so far the most effective and economic one, as it is comprised of single-cell-based populational analysis.

By tracing the process of genome chaos using an in vitro model, it becomes clear that different types of chromosomal/nuclear abnormalities are linked by the degree of CIN, the phase of evolution, and the level of system stress and stress response. For example, cells with giant nuclei can be generated by the genome chaos process, and giant cells can be linked to micronuclei clusters and more complicated translocations. To make the situation more complicated, some transitional structures can trigger further stress responses even though these will not be survived at the end of the chaotic process. As a conclusion, it is possible that in the future, we will need to monitor evolutionary mechanisms rather than specific types of chromosomal abnormalities as they are constantly changing.

Nevertheless, before we achieve the future goal of using quantitative general biomarkers (rather than using one specific type of abnormalities alone), further characterization and classification of types of abnormalities are needed, as many of them involve different names, and some confusion about them exists as well. For example, despite their similar morphological features, C-Frag differs from PCC (premature chromosome condensation), both from a morphological and mechanistic point of view (for more details, please see Stevens and Heng [2013]). Similarly, many terms are overlapping, such as chromosome pulverization, shattering, and mitotic catastrophe. These can all be termed as forms of C-Frag, a means of mitotic cell death. More generally, they are unified by genome chaos. Clearly, one important concept is the heterogeneity of cell death (Stevens et al. 2013). Drastically altered chromosomal morphological features do not mean the elimination of the system but the emergence of a new system, albeit at very low frequencies (Fig. 6.6).

The Evolutionary Mechanism of Stochastic Chromosome/ Nuclear Alterations

Prior to recent evolutionary mechanism-focused research, most chromosomal/ nuclear abnormalities are studied by different investigators within the premise of studying specific molecular mechanisms. For example, aneuploidy has mainly been

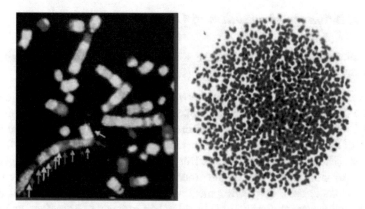

Fig. 6.6 Examples of structural and numerical chaotic genomes. Despite that there are many sub-types of chaotic genomes, structural chaotic genomes commonly involve multiple translocations (as in the SKY image, in which the chromosomes in the left corner are formed by at least 15 large chromosome fragments, some of which are indicated by arrows with different colors) (left image). On the other hand, numerical chaotic genomes can contain hundreds of chromosomes, as exemplified by the right image, in which the genome contains over 700 human chromosomes or > 15 n of DNA content. Two images are reused from Heng (2013) and Liu et al. (2014)

linked to the chromosome segregation mechanism. With various large scale -omics studies, however, many different specific molecular mechanisms have been linked to aneuploidy, which makes aneuploidy research much more complicated. This situation calls for a new strategy of studying the general evolutionary mechanisms of aneuploidy which can unify diverse molecular mechanisms (Ye et al. 2018a, b). Obviously, such a strategy should be used for studying all types of chromosomal/nuclear abnormalities (Heng 2015, 2019).

The General Causative Factor of Genome Alterations

Even though many different molecular mechanisms can be linked to a given type of abnormality (e.g., over a dozen different treatments/mechanisms can be linked to C-Frag) (Stevens et al. 2011), the general causative factors can be described as internal genomic stochasticity and stress response-mediated cellular adaptation, in addition to bio-errors produced under dynamic environmental conditions. It is important to point out that even the process of cell death can eliminate many unwanted cells (to reduce the average population size); under many circumstances, the process itself can trigger further system changes with unexpected consequences (such as the creation and/or favoring of some outliers which provide resistance). The long-term consequences, for better or worse, depend on the multiple levels of the systems and the fate of evolutionary selection.

The Evolutionary Mechanism of Genome Alterations

(a) Promoting genomic variants at the somatic cell level: solving the conflicts of constraint (germline) and dynamics (somatic)

In working to solve the conflict between species' genomic stability and the genomic dynamism necessary for adaptation (the two faces of the coin that are essential for evolution), it was realized that genome integrity is maintained by the stability of the genomic landscape of the germline (which is ensured by the function of the sex) (Heng 2007b; Gorelick and Heng 2011; Heng 2015, 2019). The genomic dynamics of the somatic cell, on the other hand, are achieved by the fuzzy inheritance of somatic cells and environmental interaction (which is promoted by the needs of cellular adaptation within changing environments). Therefore, as long as the germline's karyotype coding is preserved, somatic alterations can be pushed to very high levels. As the trade-off for the benefit of cellular adaptation, there are many disease conditions caused by the increased variants generated (Heng et al. 2016a, b; Heng 2017b).

Interestingly, the concept of system inheritance, combined with the separation of germline constraint and somatic dynamics, can also explain part of the missing heritability (Heng 2010, 2019). The gene-centric concept will not able to identify the missing heritability. Unfortunately, current major efforts are still within the genome centric framework, although they are making greater use of computational models.

(b) Genome reorganization and evolutionary potential

With so many different types of unclassified chromosomal abnormalities, and even due to the presence of just one given type, there are high degrees of morphological heterogeneity, which makes it rather challenging to understand the main function of these abnormalities. As different types of chromosomal abnormalities can be linked to many different molecular mechanisms, molecular mechanistic understanding as a whole becomes less certain. As a result, even though increased molecular knowledge is available, much of this knowledge can only explain limited cases. Examples can be found in aneuploidy and micronuclei research (Ye et al. 2018b, 2019a). As a result, the underlying common principles that can unify all of these chromosomal and nuclear variants are lacking, and the incidence of clinical prediction based on individual molecular mechanisms is low.

Clearly, a correct approach is to go above the individual molecular mechanisms (as there are so many) to search for an evolutionary and informational mechanism, which is applicable to all chromosomal abnormalities.

One holistic understanding is that regardless of their morphological and mechanistic differences, all of these NCCAs are simply chromosomal or nuclear variants with altered chromosomal codes. In other words, their informational meaning and evolutionary mechanism is the same: the creation of a new information package with evolutionary potential.

A general model has been proposed when discussing the mechanism of how genome chaos leads to a new system by reorganizing the chromosomes (Heng et al.

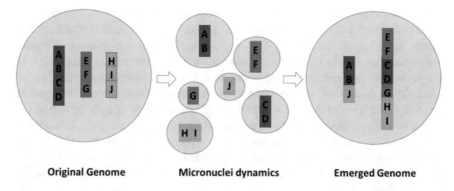

Original Genome **Micronuclei dynamics** **Emerged Genome**

Fig. 6.7 The diagram of how micronuclei create a new genome by reorganizing karyotype coding. When under a high level of stress (either internal or environmental), the cluster of micronuclei is formed, which can lead to death, proportional survival (partial population survival without altering the genome), the formation of an emergent genome through a fusion/fission cycle, or simply the combination of micronuclei with other nuclei, resulting in a new cell with an emergent genome (defined by altered chromosomal coding). (Reused from Ye et al. 2019a)

2011a). This model was also applied to explain how micronuclei clusters can form different genomes (Ye et al. 2019a). (Fig. 6.7, Micronuclei cluster model of the reorganizing of the genome)

This model can be applied to explain how different chromosomal/nuclear abnormalities contribute to new genome formation, including polyploidy/aneuploidy, sticky chromosomes, giant nuclei, and entosis (Ye et al. 2019b). All of these are associated with the stress response and unstable genome status, in conjunction with system adaptation and survival. Fundamentally, they all contribute to the emergence of an end product with altered genomic coding.

(c) Heterogeneity of abnormalities caused by fuzzy inheritance and dynamic environments

Of course, fuzzy inheritance at the chromosomal level represents the basis for the heterogeneity of chromosomal abnormalities. Fuzzy coding is responsible for the potential phenotype, and it is the environment that selects the specific phenotypes. However, the selected phenotypes can easily be altered again under different selective conditions as the inherited code itself is highly flexible, and the phenotypes themselves exist within a range of potential options, a concept which differs from classical genetic frameworks (Heng 2015, 2019; Ye et al. 2018a, b). Nature has beautifully solved the key conflict of survival as a species (by not changing the entire system) and while rendering the species' bio-information flexible enough to adapt to current conditions. Clearly, the fuzzy inheritance of somatic cells, including the separation of germline and somatic cells, plays an important role.

It should be pointed out that there is emerging interest in somatic mosaicism (Yurov et al. 2007; Iourov et al. 2008, 2010, 2019; Biesecker and Spinner 2013; Heng et al. 2013a, b) and core genomes-associated multiple levels of genomic interactions (Heng et al. 2013a, b, 2016a; Shapiro 2017, 2019; Heng 2019), which are closely related to fuzzy inheritance and genome-based evolution. These mechanisms, including minimal genomic variations in the germline, somatic alteration and mosaicism, and the host microbiome, allow diverse variants to be achieved by the same core genome interacting with other genomic and environmental factors. Under many conditions, such genome level interaction plus epigenetic changes can provide enough variations without relying on the changing of gene mutation frequencies within a population, the key mechanism of natural selection. Just passing the core genome is sufficient for passing the potential of different combinations of genomic interaction. As long as such interaction is there, there is no need to accumulate gene mutation for most traits as the environments are constantly changing back and forth.

Future Perspectives

In recent years, there have been increased reports on the significance of using various chromosomal/nuclear abnormalities in both genomic research and clinical implications (Chandrakasan et al. 2011; Heng et al. 2013a, b; Stepanenko and Kavsan 2014; Stepanenko and Dmitrenko 2015a, b; Niederwieser et al. 2016; Bloomfield and Duesberg 2016; Stepanenko and Heng 2017; Poot 2017; Rangel et al. 2017; Iourov et al. 2019; Vargas-Rondón et al. 2017; Liu et al. 2018; Heng et al. 2018; Frias et al. 2019; Ramos et al. 2018; Chin et al. 2018; Salmina et al. 2019). With an appreciation of the importance of karyotype or chromosomal coding, and of how these stochastic abnormalities can play a key role in somatic evolution, a new wave of studies will likely soon come of age. Along with some frequently discussed perspectives (Heng et al. 2016a, 2018; Heng 2013, 2015, 2019; Heng and Regan 2018; Ye et al. 2018a, 2019a, b), several issues should be addressed for further classifying and applying the knowledge of chromosomal abnormalities in clinic settings. First, the baselines of some major types of abnormalities in normal individuals and in patients are needed to be established and give reference to age, gender, and possible racial difference. Of course, for many common and complex diseases or illnesses, research is needed to examine if elevated levels of NCCAs are involved. Second, a quantitative measurement based on total chromosomal abnormalities is needed to link to different types of diseases, treatments, and overall system instability. Such studies might lead to new biomarkers based on the pattern of genome dynamics. The possibility of combining chromosomal and nuclear abnormalities together to predict system instability and evolutionary potential should also be studied. Third, the pattern of chromosomal abnormalities should be used to study the behavior of outliers within different phases of somatic evolution. The profile of outlier versus average is particularly interesting during phase transitions (Heng 2015, 2019). Fourth, another challenge is to integrate different types of variants into somatic chromosomal mosaicism (Iourov et al. 2019). Obviously, mosaicism plays an important role during the emergence of systems behavior (Heng et al. 2019).

Lastly, it should be noticed that the concept of chromosomal coding mainly applies to eukaryotes with typical chromosomes. As the chromosome represents a major innovation of our evolutionary history, the function of chromosome-based genomes drastically differs from that of prokaryotic genomes. As soon as chromosomes were formed on Earth, prokaryotes and eukaryotes have followed different games of evolution. For example, meiosis has become a main constraint for maintaining species' identities, while the breakage of chromosomal coding has become the major tool for rapid macroevolution, with increased system complexity. The chromosome-based information package has likely provided the separation of germline and somatic cells, which further increased the power of fuzzy inheritance. Of course, more research is needed to compare the evolutionary and informational mechanism of non-chromosome-based and chromosome-based genomes.

Acknowledgments This manuscript is part of our series of publications on the subject of "the mechanisms of cancer and organismal evolution." This work was partially supported by the start-up fund for Christine J. Ye from the University of Michigan's Department of Internal Medicine, Hematology/Oncology Division. We thank Eric Heng for figure preparations.

References

Baca SC, Prandi D, Lawrence MS et al (2013) Punctuated evolution of prostate cancer genomes. Cell 153(3):666–677

Bak AL, Bak P, Zeuthen J (1979) Higher levels of organization in chromosomes. J Theor Biol 76:205–217

Biesecker LG, Spinner NB (2013) A genomic view of mosaicism and human disease. Nat Rev Genet 14(5):307–320. https://doi.org/10.1038/nrg3424

Bloomfield M, Duesberg P (2016) Inherent variability of cancer-specific aneuploidy generates metastases. Mol Cytogenet 9:90

Caspersson T, Zech L, Johansson C (1970) Differential banding of alkylating fluorochromes in human chromosomes. Exp Cell Res 60:315–319

Chandrakasan S, Ye CJ, Chitlur M, Mohamed AN, Rabah R, Konski A, Heng HH, Savaşan S (2011) Malignant fibrous histiocytoma two years after autologous stem cell transplant for Hodgkin lymphoma: evidence for genomic instability. Pediatr Blood Cancer 56(7):1143–1145

Chen J, Niu N, Zhang J, Qi L, Shen W, Donkena KV, Feng Z, Liu J (2018) Polyploid giant cancer cells (PGCCs): the evil roots of cancer. Curr Cancer Drug Targets 18:1–8

Chin TF, Ibahim K, Thirunavakarasu T, Azanan MS, Lixian O, Lum SH, Yap TY, Ariffin H (2018) Nonclonal chromosomal aberrations in childhood leukemia survivors. Fetal Pediatr Pathol 37:243–253

Crasta K, Ganem NJ, Dagher R et al (2012) DNA breaks and chromosome pulverization from errors in mitosis. Nature 482(7383):53–58

Dong Z, Wang H, Chen H, Jiang H, Yuan J, Yang Z, Wang WJ, Xu F, Guo X, Cao Y, Zhu Z, Geng C, Cheung WC, Kwok YK, Yang H, Leung TY, Morton CC, Cheung SW, Choy KW (2018) Identification of balanced chromosomal rearrangements previously unknown among participants in the 1000 Genomes Project: implications for interpretation of structural variation in genomes and the future of clinical cytogenetics. Genet Med 20(7):697–707. https://doi.org/10.1038/gim.2017.170

Erenpreisa J, Kalejs M, Ianzini F, Kosmacek EA, Mackey MA, Emzinsh D et al (2005) Segregation of genomes in polyploid tumour cells following mitotic catastrophe. Cell Biol Int 29(12):1005–11.89

Forment JV, Kaidi A, Jackson SP (2012) Chromothripsis and cancer: causes and consequences of chromosome shattering. Nat Rev Cancer 12(10):663–670

Frias S, Ramos S, Salas C, Molina B, Sánchez S, Rivera-Luna R (2019) Nonclonal chromosome aberrations and genome chaos in somatic and germ cells from patients and survivors of hodgkin lymphoma. Genes (Basel) 10:37

Gorelick R, Heng HH (2011) Sex reduces genetic variation: a multidisciplinary review. Evolution 65(4):1088–1098

Haaf T, Schmid M (1989) 5-Azadeoxycytidine induced undercondensation in the giant X chromosomes of Microtus agrestis. Chromosoma 98(2):93–98

Heng HH (2000) Released chromatin or DNA fiber preparations for high-resolution fiber FISH. Methods Mol Biol 123:69–81

Heng HH (2007a) Cancer genome sequencing: the challenges ahead. BioEssays 29(8):783–794

Heng HH (2007b) Elimination of altered karyotypes by sexual reproduction preserves species identity. Genome 50(5):517–524. https://doi.org/10.1139/G07-039

Heng HH (2007c) Karyotypic chaos, a form of non-clonal chromosome aberrations, plays a key role for cancer progression and drug resistance. FASEB: Nuclear Structure and Cancer. Vermont Academy, Saxtons River, Vermont, 2007

Heng HH (2009) The genome-centric concept: resynthesis of evolutionary theory. BioEssays 31:512–525

Heng HH (2010) Missing heritability and stochastic genome alterations. Nat Rev Genet 11(11):813

Heng HH (2013) Genomics: HeLa genome versus donor's genome. Nature 501:167

Heng HH (2015) Debating cancer: the paradox in cancer research. World Scientific Publishing Co., Singapore. ISBN:978-981-4520-84-3

Heng HH (2017a) Chapter 5 – The genomic landscape of cancers. In: Ujvari B, Roche B, Thomas F (eds) Ecology and evolution of cancer. Academic, pp 69–86

Heng HH (2017b) Heterogeneity-mediated cellular adaptation and its trade-off: searching for the general principles of diseases. J Eval Clin Pract 23(1):233e237. https://doi.org/10.1111/jep.12598

Heng HH (2019) Genome chaos: rethinking genetics, evolution, and molecular medicine. Academic, Cambridge, MA. ISBN:978-012-8136-35-5

Heng HH, Chen W (1985) The study of the chromatin and the chromosome structure for bufo gargarizans by the light microscope. J Sichuan Normal Univ Nat Sci 2:105–109

Heng HH, Regan S (2018) A systems biology perspective on molecular cytogenetics. Curr Bioinforma 12:4e10. https://doi.org/10.2174/1574893611666160606163419

Heng HH, Shi XM (1997) From free chromatin analysis to high resolution fiber FISH. Cell Res 7(1):119–124

Heng HH, Tsui LC (1994) Free chromatin mapping by FISH. Methods Mol Biol 33:109–122

Heng HH, Chen W, Wang Y (1988a) Effects of pingyanymycin on chromosomes: a possible structural basis for chromosome aberration. Mutat Res 199:199–205

Heng HH, Lin R, Zhao X et al (1988b) Structure of the chromosome and its formation. II. Studies on the sister unit fibers. Nucleus 30:2–9

Heng HH, Squire J, Tsui L (1991) Chromatin mapping – a strategy for physical characterization of the human genome by hybridization in situ. In Paper presented at the 8th Int Cong hum gen Am J hum gent, DC, USA

Heng HH, Squire J, Tsui LC (1992) High-resolution mapping of mammalian genes by in situ hybridization to free chromatin. Proc Natl Acad Sci U S A 89:9509–9513

Heng HH, Spyropoulos B, Moens PB (1997) FISH technology in chromosome and genome research. BioEssays 19(1):75–84

Heng HH, Krawetz SA, Lu W, Bremer S, Liu G, Ye CJ (2001) Re-defining the chromatin loop domain. Cytogenet Cell Genet 93(3–4):155–161

Heng HH, Ye CJ, Yang F, Ebrahim S, Liu G, Bremer SW, Thomas CM, Ye J, Chen TJ, Tuck-Muller C, Yu JW, Krawetz SA, Johnson A (2003) Analysis of marker or complex chromosomal rearrangements present in pre- and post-natal karyotypes utilizing a combination of G-banding, spectral karyotyping and fluorescence in situ hybridization. Clin Genet 63(5):358–367

Heng HH, Stevens JB, Liu G, Bremer SW, Ye CJ (2004a) Imaging genome abnormalities in cancer research. Cell Chromosome 3(1):1

Heng HH, Goetze S, Ye CJ et al (2004b) Chromatin loops are selectively anchored using scaffold/matrix-attachment regions. J Cell Sci 117(Pt 7):999e1008. https://doi.org/10.1242/jcs.00976

Heng HH, Bremer SW, Stevens J, Ye KJ, Miller F, Liu G, Ye CJ (2006a) Cancer progression by non-clonal chromosome aberrations. J Cell Biochem 98(6):1424–1435

Heng HH, Liu G, Bremer S, Ye KJ, Stevens J, Ye CJ (2006b) Clonal and non-clonal chromosome aberrations and genome variation and aberration. Genome 49(3):195–204

Heng HH, Stevens JB, Liu G, Bremer SW, Ye KJ, Reddy PV et al (2006c) Stochastic cancer progression driven by nonclonal chromosome aberrations. J Cell Physiol 208:461–472

Heng HH, Stevens JB, Lawrenson L, Liu G, Ye KJ, Bremer SW, Ye CJ (2008) Patterns of genome dynamics and cancer evolution. Cell Oncol 30:513–514

Heng HH, Bremer WS, Stevens JB, Ye KJ, Liu G, Ye CJ et al (2009) Genetic and epigenetic heterogeneity in cancer: a genome centric perspective. J Cell Physiol 220(3):538–547. https://doi.org/10.1002/jcp.21799

Heng HH, Liu G, Stevens JB, Bremer SW, Ye KJ, Abdallah BY et al (2011a) Decoding the genome beyond sequencing: the next phase of genomic research. Genomics 98(4):242–252. https://doi.org/10.1016/j.ygeno.2011.05.008

Heng HH, Stevens JB, Bremer SW, Liu G, Abdallah BY, Ye CJ (2011b) Evolutionary mechanisms and diversity in cancer. Adv Cancer Res 112:217–253. https://doi.org/10.1016/B978-0-12-387688-1.00008-9

Heng HH, Liu G, Stevens JB, Abdallah BY, Horne SD, Ye KJ et al (2013a) Karyotype heterogeneity and unclassified chromosomal abnormalities. Cytogenet Genome Res 139(3):144–157. https://doi.org/10.1159/000348682.Heng

Heng HH, Bremer SW, Stevens JB, Horne SD, Liu G, Abdallah BY et al (2013b) Chromosomal instability (CIN): what it is and why it is crucial to cancer evolution. Cancer Metastasis Rev 32:325–340

Heng HH, Regan SM, Liu G et al (2016a) Why it is crucial to analyze non clonal chromosome aberrations or NCCAs? Mol Cytogenet 9:15. https://doi.org/10.1186/s13039-016-0223-2

Heng HH, Regan S, Ye C (2016b) Genotype, environment, and evolutionary mechanism of diseases. Environ Dis 1:14e23

Heng HH, Horne SD, Chaudhry S, Regan SM, Liu G, Abdallah BY, Ye CJ (2018) A Postgenomic Perspective on Molecular Cytogenetics. Curr Genomics 19:227–239

Heng HH, Liu G, Alemara S, Regan S, Armstrong Z, Ye CJ (2019) The mechanisms of how genomic heterogeneity impacts bio-emergent properties: the challenges for precision medicine. In embracing complexity in health. In: Sturmberg J (ed) Embracing complexity in health. Springer, Cham, pp 95–109. https://doi.org/10.1007/978-3-030-10940-0_6

Horne SD, Heng HH (2014) Genome chaos, chromothripsis and cancer evolution. J Cancer Stud Ther 1:1–6

Horne SD, Stevens JB, Abdallah BY, Liu G, Bremer SW, Ye CJ, Heng HH (2013) Why imatinib remains an exception of cancer research. J Cell Physiol 228(4):665–670

Horne SD, Chowdhury SK, Heng HH (2014) Stress, genomic adaptation, and the evolutionary trade-off. Front Genet 5:92. https://doi.org/10.3389/fgene.2014.00092

Horne SD, Pollick SA, Heng HH (2015) Evolutionary mechanism unifies the hallmarks of cancer. Int J Cancer 136(9):2012–2021

Iafrate AJ, Feuk L, Rivera MN, Listewnik ML, Donahoe PK, Qi Y et al (2004) Detection of large-scale variation in the human genome. Nat Genet 36:949–951

Iourov IY, Vorsanova SG, Yurov YB (2008) Chromosomal mosaicism goes global. Mol Cytogenet 1:26

Iourov IY, Vorsanova SG, Yurov YB (2010) Somatic genome variations in health and disease. Curr Genomics 11:387–396

Iourov IY, Vorsanova SG, Yurov YB, Kutsev SI (2019) Ontogenetic and pathogenetic views on somatic chromosomal mosaicism. Genes (Basel) 10(5). https://doi.org/10.3390/genes10050379

Langer PR, Waldrop AA, Ward DC (1981) Enzymatic synthesis of biotin-labelled polynucleotides: novel nucleic acid affinity probes. Proc Natl Acad Sci U S A 78:6633–6637

Lejeune J, Marie G, Turpin R (1959) Les chromosomeshumains en culture de tissus. CR Hebd Séances Acad Sci (Paris) 248:602–603

Lichter P, Tang CC, Call CK et al (1990) High resolution mapping of human chromosome 11 by in situ hybridisation with cosmid clones. Science 247:64–69

Liehr T (2016) Cytogenetically visible copy number variations (CG-CNVs) in banding and molecular cytogenetics of human; about heteromorphisms and euchromatic variants. Mol Cytogenet 22:9–5

Liu P, Erez A, Nagamani SC et al (2011) Chromosome catastrophes involve replication mechanisms generating complex genomic rearrangements. Cell 146(6):889–903

Liu G, Stevens JB, Horne SD, Abdallah BY, Ye KJ, Bremer SW, Ye CJ, Chen DJ, Heng HH (2014) Genome chaos: survival strategy during crisis. Cell Cycle 13(4):528–537

Liu G, Ye CJ, Chowdhury SK, Abdallah BY, Horne SD, Nichols D, Heng HH (2018) Detecting chromosome condensation defects in Gulf war illness patients. Curr Genomics 19(3):200–206. https://doi.org/10.2174/1389202918666170705150819

Mitelman F (2000) Recurrent chromosome aberrations in cancer. Mutat Res 462(2e3):247–253

Niederwieser C, Nicolet D, Carroll AJ, Kolitz JE, Powell BL, Kohlschmidt J, Stone RM, Byrd JC, Mrózek K, Bloomfield CD (2016) Chromosome abnormalities at onset of complete remission are associated with worse outcome in patients with acute myeloid leukemia and an abnormal karyotype at diagnosis: CALGB 8461 (Alliance). Haematologica 101:1516–1523

Nowell PC, Hungerford DA (1960) A minute chromosome in human chronic myelocytic leukaemia. Science 132:1497

Poot M (2017) Of simple and complex genome rearrangements, chromothripsis, chromoanasynthesis, and chromosome chaos. Mol Syndromol 8(3):115–117

Ramos S, Navarrete-Meneses P, Molina B, Cervantes-Barragán DE, Lozano V, Gallardo E, Marchetti F, Frias S (2018) Genomic chaos in peripheral blood lymphocytes of Hodgkin's lymphoma patients one year after ABVD chemotherapy/radiotherapy. Environ Mol Mutagen 59(8):755–768. https://doi.org/10.1002/em.22216

Rangel N, Forero-Castro M, Rondón-Lagos M (2017) New insights in the cytogenetic practice: karyotypic chaos, non-clonal chromosomal alterations and chromosomal instability in human cancer and therapy response. Genes 8:155

Righolt C, Mai S (2012) Shattered and stitched chromosomes-chromothripsis and chromoanasynthesis-manifestations of a new chromosome crisis? Genes Chromosomes Cancer 51(11):975–981

Rowley JD (2013) Genetics. A story of swapped ends. Science 340(6139):1412–1413. https://doi.org/10.1126/science.1241318

Salmina K, Huna A, Kalejs M, Pjanova D, Scherthan H, Cragg M et al (2019) The cancer aneuploidy paradox: in the light of evolution. Genes 10(2):83. https://doi.org/10.3390/genes10020083

Schröck E, du Manoir S, Veldman T, Schoell B, Wienberg J, Ferguson-Smith MA, Ning Y, Ledbetter DH, Bar-Am I, Soenksen D, Garini Y, Ried T (1996) Multicolor spectral karyotyping of human chromosomes. Science 273(5274):494–497

Setlur SR, Lee C (2012) Tumor archaeology reveals that mutations love company. Cell 149(9):959–961

Shapiro JA (2017) Living organisms author their read-write genomes in evolution. Biology (Basel) 6(4). https://doi.org/10.3390/biology6040042

Shapiro JA (2019) No genome is an island: toward a 21st century agenda for evolution. Ann N Y Acad Sci 1447(1):21–52. https://doi.org/10.1111/nyas.14044

Sheltzer JM, Blank HM, Pfau SJ, Tange Y, George BM, Humpton TJ, Brito IL, Hiraoka Y, Niwa O, Amon A (2011) Aneuploidy drives genomic instability in yeast. Science 333(6045):1026–1030. https://doi.org/10.1126/science.1206412

Siegel JJ, Amon A (2012) New insights into the troubles of aneuploidy. Annu Rev Cell Dev Biol 28:189–214

Smith L, Plug A, Thayer M (2001) Delayed replication timing leads to delayed mitotic chromosome condensation and chromosomal instability of chromosome translocations. Proc Natl Acad Sci U S A 98(23):13300–13305

Speicher MR, Gwyn Ballard S, Ward DC (1996) Karyotyping human chromosomes by combinatorial multi-fluor FISH. Nat Genet 12(4):368–375

Stepanenko AA, Dmitrenko VV (2015a) HEK293 in cell biology and cancer research: phenotype, karyotype, tumorigenicity, and stress-induced genome-phenotype evolution. Gene 569(2):182–190. https://doi.org/10.1016/j.gene.2015.05.065

Stepanenko AA, Dmitrenko VV (2015b) Pitfalls of the MTT assay: direct and off-target effects of inhibitors can result in over/underestimation of cell viability. Gene 574(2):193–203. https://doi.org/10.1016/j.gene.2015.08.009

Stepanenko AA, Heng HH (2017) Transient and stable vector transfection: pitfalls, off-target effects, artifacts. Mutat Res 773:91–103

Stepanenko AA, Kavsan VM (2014) Karyotypically distinct U251, U373, and SNB19 glioma cell lines are of the same origin but have different drug treatment sensitivities. Gene 540(2):263–265. https://doi.org/10.1016/j.gene.2014.02.053

Stephens PJ, Greenman CD, Fu B et al (2011) Massive genomic rearrangement acquired in a single catastrophic event during cancer development. Cell 144(1):27–40

Stevens JB, Heng HH (2013) Differentiating chromosome fragmentation and premature chromosome condensation. In: Yurov Y, Vorsanova S, Iourov I (eds) Human interphase chromosomes. Springer, New York

Stevens JB, Liu G, Bremer SW, Ye KJ, Xu W, Xu J, Sun Y, Wu GS, Savasan S, Krawetz SA, Ye CJ, Heng HH (2007) Mitotic cell death by chromosome fragmentation. Cancer Res 67(16):7686–7694

Stevens JB, Abdallah BY, Liu G, Ye CJ, Horne SD, Wang G, Savasan S, Shekhar M, Krawetz SA, Hüttemann M, Tainsky MA, Wu GS, Xie Y, Zhang K, Heng HH (2011) Diverse system stresses: common mechanisms of chromosome fragmentation. Cell Death Dis 2:e178. https://doi.org/10.1038/cddis.2011.60

Stevens JB, Horne SD, Abdallah BY, Ye CJ, Heng HH (2013) Chromosomal instability and transcriptome dynamics in cancer. Cancer Metastasis Rev 32(3–4):391–402. https://doi.org/10.1007/s10555-013-9428-6

Stevens JB, Liu G, Abdallah BY, Horne SD, Ye KJ, Bremer SW, Ye CJ, Krawetz SA, Heng HH (2014) Unstable genomes elevate transcriptome dynamics. Int J Cancer 134(9):2074–2087

Tjio J-H, Levan A (1956) The chromosome number of man. Hereditas 42:1–6

Vargas-Rondón N, Villegas VE, Rondón-Lagos M (2017) The role of chromosomal instability in cancer and therapeutic responses. Cancers 10:4

Walen KH (2005) Budded karyoplasts from multinucleated fibroblast cells contain centrosomes and change their morphology to mitotic cells. Cell Biol Int 29(12):1057–65.94

Ye CJ, Stevens JB, Liu G, Ye KJ, Yang F, Bremer SW, Heng HH (2006) Combined multicolor-FISH and immunostaining. Cytogenet Genome Res 114(3–4):227–234

Ye CJ, Regan S, Liu G, Alemara S, Heng HH (2018a) Understanding aneuploidy in cancer through the lens of system inheritance, fuzzy inheritance and emergence of new genome systems. Mol Cytogenet 11:31. https://doi.org/10.1186/s13039-018-0376-2

Ye CJ, Liu G, Heng HH (2018b) Experimental induction of genome chaos. Methods Mol Biol 1769:337–352. https://doi.org/10.1007/978-1-4939-7780-2_21

Ye CJ, Sharpe Z, Alemara S, Mackenzie S, Liu G, Abdallah B, Horne S, Regan S, Heng HH (2019a) Micronuclei and genome chaos: changing the system inheritance. Genes (Basel) 10(5). https://doi.org/10.3390/genes10050366

Ye CJ, Stilgenbauer L, Moy A, Liu G, Heng HH (2019b) What is karyotype coding and why is genomic topology important for cancer and evolution? Front Genet. https://doi.org/10.3389/fgene.2019.01082

Yurov YB, Iourov IY, Vorsanova SG, Liehr T, Kolotii AD, Kutsev SI, Pellestor F, Beresheva AK, Demidova IA, Kravets VS et al (2007) Aneuploidy and confined chromosomal mosaicism in the developing human brain. PLoS One 2:e558

Zhang S, Mercado-Uribe I, Xing Z, Sun B, Kuang J, Liu J (2014) Generation of cancer stem-like cells through the formation of polyploid giant cancer cells. Oncogene 33:116–128

Zhu J, Pavelka N, Bradford WD, Rancati G, Li R (2012) Karyotypic determinants of chromosome instability in aneuploid budding yeast. PLoS Genet 8:e1002719

Chapter 7
Twenty-First Century FISH: Focus on Interphase Chromosomes

Svetlana G. Vorsanova, Yuri B. Yurov, Oxana S. Kurinnaia, Alexei D. Kolotii, and Ivan Y. Iourov

Abstract Interphase molecular cytogenetics provides opportunities for analysis of chromosomes in almost all types of human cells at any stage of the cell cycle. Generally, interphase fluorescence in situ hybridization (I-FISH) is a basic technological platform for visualization of individual chromosomes (chromosomal regions) in single cells. The achievements of studying human interphase chromosomes have allowed numerous discoveries in chromosome research (molecular cytogenetics) and genomics (cytogenomics). In the postgenomic era, interphase chromosome analysis by I-FISH remains an important part of biomedical research. Here, we describe the spectrum of FISH applications with special emphasis on interphase chromosome biology and molecular cytogenetic/cytogenomic diagnosis.

Introduction

Fluorescence in situ hybridization (FISH) is recognized as one of essential technological platforms for molecular cytogenetics. During the last decades, FISH has been found useful for a wide spectrum of applications from molecular diagnosis to basic chromosome biology (van der Ploeg 2000; Vorsanova et al. 2010c; Yurov et al. 2013; Liehr 2017; Hu et al. 2020). Previous edition of this book contained a chapter

S. G. Vorsanova · Y. B. Yurov · O. S. Kurinnaia · A. D. Kolotii
Veltischev Research and Clinical Institute for Pediatrics, Pirogov Russian National Research Medical University, Ministry of Health of Russian Federation, Moscow, Russia

Mental Health Research Center, Russian Ministry of Science and Higher Education, Moscow, Russia

I. Y. Iourov (✉)
Mental Health Research Center, Moscow, Russia

Veltischev Research and Clinical Institute for Pediatrics of the Pirogov Russian National Research Medical University, Ministry of Health of Russian Federation, Moscow, Russia

Medical Genetics Department of Russian Medical Academy of Continuous Postgraduate Education, Moscow, Russia

© Springer Nature Switzerland AG 2020
I. Iourov et al. (eds.), *Human Interphase Chromosomes*,
https://doi.org/10.1007/978-3-030-62532-0_7

dedicated to technological solutions in interphase chromosome biology, i.e., interphase FISH (I-FISH) (Vorsanova et al. 2013). Since that time, no groundbreaking technological developments have been made in I-FISH or related techniques for studying interphase chromosomes. However, it seems that reconsidering technological aspects of interphase molecular cytogenetics is required, inasmuch as general decrease of interest to molecular cytogenetics (e.g., FISH) may be observed in the postgenomic era (Liehr 2017; Iourov 2019b; Heng 2020). Here we have reviewed I-FISH in the light of its application in the postgenomic context.

No fewer than one million cytogenetic and molecular cytogenetic analyses are suggested to be performed per year (Gersen and Keagle 2005). Molecular (cytogenetic) diagnosis is the standard of medical care for clinical genetics, reproduction, oncology, neurology, psychiatry, etc. (Vorsanova et al. 2010d; Bint et al. 2013; Liehr et al. 2015; Viotti 2020). The diagnostic value of FISH has been repeatedly noted and has been considered as either an alternative to conventional cytogenetic analysis or a confirmatory method (Feuk et al. 2006; Iourov et al. 2008c; Martin and Warburton 2015; Liehr 2017). In addition, I-FISH-like protocols are used in microbiology (Frickmann et al. 2017), genetic toxicology (Hovhannisyan 2010; Iurov et al. 2011), somatic cell genetics/genomics (Yurov et al. 2001, 2018b, 2019a; Iourov et al. 2008b, 2010b), aging research (Yurov et al. 2009, 2010a), and single-cell biology (Iourov et al. 2012, 2013a; Yurov et al. 2019b; Gupta et al. 2020). In summary, one can be certain that FISH-based molecular cytogenetic analysis has an important role in biomedicine.

In basic research, I-FISH is used for studying somatic chromosomal mosaicism (Iourov et al. 2006c, 2010a, 2017, 2019a, d; Arendt et al. 2009; Bakker et al. 2015; Andriani et al. 2019) and genome organization in interphase nuclei at the chromosomal level (Rouquette et al. 2010; Iourov 2012; Cui et al. 2016; Baumgartner et al. 2018). A successful study of the aforementioned phenomena requires the application of various I-FISH-based techniques, which are described in this chapter.

I-FISH

FISH is an umbrella term for molecular cytogenetic visualization techniques for studies of genome (specific genomic loci) using DNA/RNA probes. FISH resolution is defined by DNA sequence size of the probes. DNA probes are centromeric and telomeric (repetitive-sequence DNA), site-specific (euchromatic DNA, e.g., gene DNAs), and whole chromosome painting (wcp; hybridizing to the whole chromosomes DNAs) (Liehr et al. 2004; Iourov et al. 2008b; Vorsanova et al. 2013). Basically, I-FISH requires (i) cell suspensions prepared specifically for FISH analysis, (ii) denaturation of chromosomal DNA and hybridization, and (iii) microscopic visual and digital analysis of FISH results (Iourov et al. 2006b, 2017; Yurov et al. 2017).

FISH analysis of repetitive genomic sequences is performed with centromeric (chromosome enumeration or chromosome-specific). I-FISH with DNA probes for repetitive sequences is applicable for analysis of nuclear chromosomal organization and numerical chromosome abnormalities (Yurov et al. 1996; Soloviev et al. 1998). I-FISH using centromeric DNA probes is used in molecular diagnosis (medical genetics, oncology, and reproduction) (Pinkel et al. 1986; Vorsanova et al. 1986, 2005b, 2010a; Yurov et al. 2007b, 2010b; Savic and Bubendorf 2016). Furthermore, I-FISH demonstrates these protocols highly applicable for studies encompassing chromosome biology, genome research (chromosomal and nuclear), evolution, behavior, and variation in health and disease (Liehr 2017). Near 100% hybridization efficiency and chromosome specificity (apart from chromosomes 5 and 19, 13 and 21, 14 and 22) defines I-FISH with these DNA probes as an effective molecular cytogenetic approach (e.g., analysis of homologous chromosomes in interphase) (Iourov et al. 2006d; Wan 2017; Russo et al. 2016; Yurov et al. 2017; Weise et al. 2019) (Fig. 7.1). I-FISH is shown to have the highest efficiency in uncovering mosaicism rates (Iourov et al. 2013b).

Site-specific DNA probes (yeast artificial chromosomes or YACS, bacterial artificial chromosomes or BACs, P1-derived artificial chromosomes or PACs, cosmids) provide the visualization of euchromatic chromosomal DNA. These probes are useful for targeted FISH assays to diagnose structural and, more rarely, numerical chromosome imbalances (Fig. 7.2) (Soloviev et al. 1995; Liehr et al. 2004; Riegel 2014; Cheng et al. 2017; Liehr 2017). The use of I-FISH assays with site-specific DNA probes is systematically applied in cancer research and molecular oncologic diagnosis (Chrzanowska et al. 2020). In the postgenomic era, these methods is applicable for mapping altered genomic loci, chromosome instability analysis, and arrangement of specific chromosomal loci in interphase.

I-FISH with chromosome-enumeration and site-specific probes may be affected by several phenomena occurring in interphase nuclei. Variable efficiency of hybridization complicates simultaneous applications of different probe sets, i.e., some signals can be invisible because of intensity differences (Iourov et al. 2006a). S phase DNA replication cause doubling of I-FISH signals (site-specific and centromeric probes) (Soloviev et al. 1995; Vorsanova et al. 2001a). False-positive chromosome abnormalities may be "uncovered" due to specific nuclear interphase chromosome architecture (genome organization). For instance, chromosomal associations affect I-FISH interpretation. Chromosomal associations/pairing are common in postmitotic cells types (Yurov et al. 2005, 2007a, 2008, 2014, 2018a; Iourov et al. 2009a, b). Quantitative FISH (QFISH) is used to differ between chromosome losses and chromosomal associations (discussed below). Solutions for these problems are given in Fig. 7.3. Finally, an appreciable increase of FISH efficiency may be achieved using microwave activation (for more details, see Soloviev et al. [1994], Durm et al. [1997], Weise et al. [2005]).

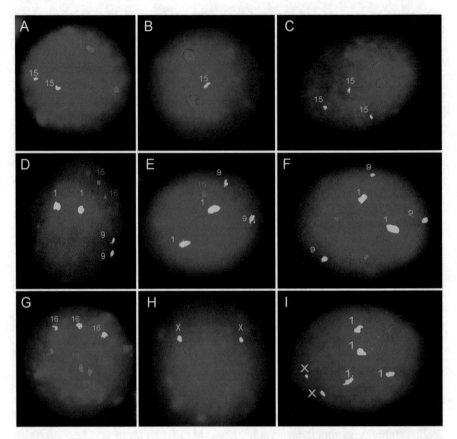

Fig. 7.1 Two- and three-color I-FISH with centromeric DNA probes. (**a**) Normal diploid nucleus with two signals for chromosome 1 and chromosome 15. (**b**) Monosomic nucleus with two signals for chromosome 1 and one signal for chromosome 15. (**c**) Trisomic nucleus with two signals for chromosome 1 and three signals for chromosome 15. (**d**) Normal diploid nucleus with two signals for chromosome 1, chromosome 9, and chromosome 16. (**e**) Monosomic nucleus with two signals for chromosome 1 and chromosome 9 and one signal for chromosome 16. (**f**) Trisomic nucleus with two signals for chromosome 1 and chromosome 16 and three signals for chromosome 9. (**g**) Triploid nucleus with three signals for chromosome 16 and chromosome 18. (**h**) Tetraploid nucleus with two signals for chromosome X and chromosome Y. (**i**) Tetraploid nucleus with two signals for chromosome X and chromosome Y and four signals for chromosome 1. (Copyright © Vorsanova et al. 2010c; licensee BioMed Central Ltd. This is an Open Access article distributed under the terms of the Creative Commons Attribution License, http://creativecommons.org/licenses/by/2.0)

ICS-MCB

Microdissected DNA probes may be combined to produce pseudo-G banding using FISH or multicolor banding (MCB) (Liehr et al. 2002). This technique may be applied to interphase chromosomes in a chromosome-specific manner. Interphase chromosome-specific MCB (ICS-MCB) allow the visualization of interphase

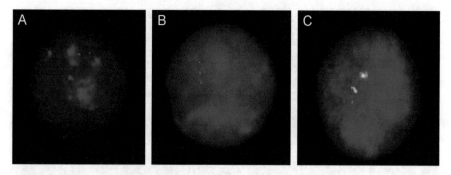

Fig. 7.2 I-FISH with site-specific DNA probes. (**a**) Normal diploid nucleus with two signals for chromosome 21. (**b**) Trisomic nucleus with three signals for chromosome 21. (**c**) Interphase nucleus exhibiting co-localization of *ABL* and *BCR* genes probably due to t(9;22)/Philadelphia chromosome. (Copyright © Vorsanova et al. 2010a; licensee BioMed Central Ltd. This is an Open Access article distributed under the terms of the Creative Commons Attribution License, http:// creativecommons.org/licenses/by/2.0)

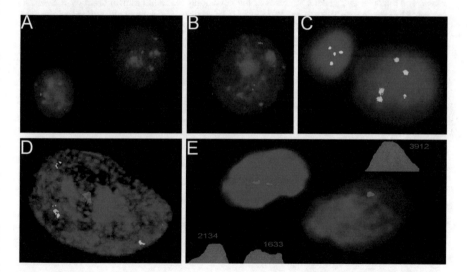

Fig. 7.3 Problems of I-FISH with centromeric/site-specific DNA probes. (**a**) and (**b**) Replication of specific genomic loci (LSI21 probe). Some nuclei exhibit replicated signals, whereas in some nuclei, it is not apparent. Note the distance between signals can be more than a diameter of a signal. (**c**) Asynchronous replication of a signal (DXZ1) in case of tetrasomy of chromosome X. Note the difficulty to make a definitive conclusion about number of signals in the right nucleus. (**d**) Two-color FISH with centromeric/site-specific DNA probes for chromosome 1 shows chromosomal associations in a nucleus isolated from the adult human brain. Note the impossibility to identify number of chromosomes. (**e**) QFISH demonstrating an association of centromeric regions of homologous chromosomes 9, but not a monosomy or chromosome loss. (Copyright © Vorsanova et al. 2010a; licensee BioMed Central Ltd. This is an Open Access article distributed under the terms of the Creative Commons Attribution License, http://creativecommons.org/licenses/by/2.0)

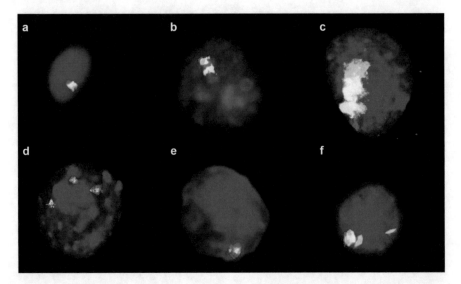

Fig. 7.4 Molecular cytogenetic analyses of the developing and adult human brain by ICS-MCB: (**a**) loss of chromosome 18 (monosomy) in a cell isolated from telencephalic regions of the fetal brain; (**b**) loss of chromosome 16 (monosomy) in a cell isolated from the cerebral cortex of the normal human brain; (**c**) loss of chromosome 1 (monosomy) in a cell isolated from the cerebral cortex of the schizophrenia brain; (**d**) gain of chromosome 21 (trisomy) in a cell isolated from the cerebral cortex of the Alzheimer's disease brain; (**e**) loss of chromosome 21 (monosomy) in a cell isolated from the cerebellum of the ataxia-telangiectasia brain; (**f**) chromosome instability in the cerebellum of the ataxia-telangiectasia brain manifesting as the presence of a rearranged chromosome 14 or der(14)(14pter- > 14q12:). (From Yurov et al. 2013 (Fig. 9.2) reproduced with permission of Springer Nature in the format reuse in a book/textbook via Copyright Clearance Center)

chromosomes in their integrity at molecular resolution (Iourov et al. 2006a, 2007). The method has been found highly effective for analysis of interphase chromosome instability and nuclear genome organization at chromosomal level (Iourov et al. 2006a, 2009a, b, 2019a; Yurov et al. 2007a, 2008, 2010b, 2014, 2019b; Liehr and Al-Rikabi 2019; Weise et al. 2019). Figure 7.4 gives a series of examples of ICS-MCB.

Immuno-FISH

Immuno-FISH is the combination of immunohistochemical detection of proteins and I-FISH (Liehr 2017). Our experience demonstrates that this technique is useful for studying chromosome instability in the human brain following by uncovering new mechanisms for neurodegeneration (Iourov et al. 2009a, b; Yurov et al. 2018b, 2019a). More precisely, immuno-FISH using NeuN antibody allows the detection of chromosomal DNA in neuronal cells (Fig. 7.5).

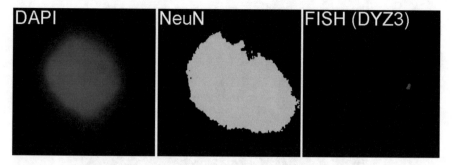

Fig. 7.5 Immuno-FISH. I-FISH using centromeric probe for chromosome Y (DYZ3) with immu-nostaining by NeuN (neuron-specific antibody) performed for the analysis of cells isolated from the human brain. (Copyright © Vorsanova et al. 2010a; licensee BioMed Central Ltd. This is an Open Access article distributed under the terms of the Creative Commons Attribution License, http://creativecommons.org/licenses/by/2.0)

QFISH

Interindividual variability of centromeric (heterochromatic) DNAs has been used of developing QFISH. This method is applicable for metaphase and interphase analy-sis of human chromosomes (Iourov et al. 2005; Vorsanova et al. 2005a; Iourov 2017). QFISH with chromosome-enumeration probes may be used for the detection of numerical imbalances of interphase chromosome (monosomy or chromosome loss). The latter is useful for prenatal and postnatal molecular diagnosis, cancer diagnosis and prognosis, and analysis of somatic genomic variability (Iourov 2017; Wan 2017; Yurov et al. 2017) (Fig. 7.6).

Molecular Diagnosis

An advantage of FISH-based techniques is referred to the availability of single-cell analysis (Iourov et al. 2012; Moffitt et al. 2016; Zhang et al. 2018). Despite the availability of DNA sequencing technologies for single-cell analysis (Knouse et al. 2014; Gawad et al. 2016), these cannot substitute FISH due to following reasons: FISH has the highest possible cell scoring potential and allows visualization of arrangement of genomic loci in interphase/metaphase chromosomes (Moffitt et al. 2016; Yurov et al. 2018b, 2019b). Accordingly, I-FISH is an important technique used in molecular cytogenetic diagnosis. Chromosomal imbalances cause a wide spectrum of diseases from congenital malformations, intellectual disability, autism, epilepsy, cancers, neurodegeneration, and reproductive problems (Vorsanova et al. 2001b, 2007, 2010b; Yurov et al. 2001, 2007b, 2019a, b; Gersen and Keagle 2005; Iourov et al. 2006c, 2008a, b, 2010b, 2011; Ye et al. 2019). Thus, the aforemen-tioned FISH methods may be applicable for the molecular diagnosis. Since a diag-nosis is aimed at uncovering molecular and cellular mechanisms for a disease, FISH

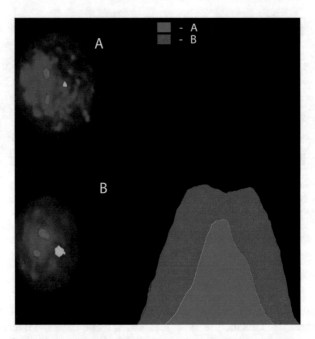

Fig. 7.6 QFISH with using enumeration-centromeric probes for chromosomes 1 (red signals/ D1Z1) and X (green signals/DXZ1): Nucleus A demonstrates a green signal with a relative intensity of 2120 pixels—true X chromosome monosomy. Nucleus B demonstrates a green signal with a relative intensity of 4800 pixels—two overlapping chromosome X signals but not a chromosome loss. (From Yurov et al. 2017 reproduced with permission of Springer Nature in the format reuse in a book/textbook via Copyright Clearance Center)

should be considered as a technique additional to whole-genome analysis (e.g., whole-genome sequencing or molecular karyotyping) for uncovering processes, which are involved in the pathogenetic cascade of a disease (i.e., chromosome instability). The postgenomic era offers numerous possibilities for pathway-based classification of genome variations to model functional consequences of a genomic change. As a result, candidate processes may be suggested (Iourov 2019b; Iourov et al. 2019b, c). Currently, several bioinformatics tools are available for molecular cytogenetics (Iourov et al. 2012, 2014b; Zeng et al. 2012). Once applied, knowledge about mechanisms of disease mediated by chromosome abnormalities allows to propose successful therapeutic strategies for presumably incurable genetic conditions (Iourov 2016; Iourov et al. 2015b). Our experience of combination of whole-genome analysis (molecular karyotyping), I-FISH, and bioinformatics analysis is shown by Fig. 7.7 (Iourov et al. 2015a). Moreover, I-FISH analysis of chromosome inability may be integrated into molecular cytogenetic diagnostic workflows (Iourov et al. 2014a).

Taking into account promising biomarkers revealed by FISH, an algorithm for identifying disease mechanisms may be proposed. To succeed, two data sets are required: (1) cytogenetic/FISH data set (analysis of large cell populations for

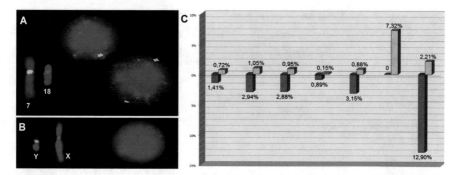

Fig. 7.7 Interphase FISH analysis of CIN (somatic aneuploidy). (**a**) FISH with DNA probes for chromosomes 7 (green) and 18 (red) showing chromosome 7 loss in the right nucleus (metaphase chromosomes show positive signals for these DNA probes). (**b**) Interphase FISH with DNA probes for chromosomes Y (green) and X (red) showing chromosome Y loss in the nucleus (metaphase chromosomes show positive signals for these DNA probes). (**c**) Rates of chromosome losses (red bars) and gains (golden bars). (From Iourov et al. 2015a, an article is distributed under the terms of the Creative Commons Attribution 4.0 International License, http://creativecommons.org/licenses/by/4.0/)

uncovering intercellular karyotypic variations) and (2) data set obtained by molecular karyotyping and analyzed using systems biology (bioinformatic) methodology for determining functional consequences of regular genomic variations. Once obtained, correlative analysis between these data sets is to be performed (Iourov 2019a; Vorsanova et al. 2019). Figure 7.8 reproduces this algorithm.

Conclusion

I-FISH seems to be an important technological part of current biomedical research and molecular diagnosis. Regardless of significant achievements in genomics and molecular biology, there is a wide spectrum of applications of this molecular cytogenetic technique. Mosaic chromosome abnormalities and chromosomal instability are relevant to numerous areas of biomedicine and require specific molecular cytogenetic approaches to the detection. Indeed, I-FISH-based techniques have to be included in the algorithms of detecting somatic genome variations at chromosomal and sub-chromosomal levels. In addition to detecting chromosomal mosaicism per se, I-FISH-based techniques are applicable to monitor somatic genomic changes and/or uncovering genome/chromosome insatiability, which may be either a cause of disease or an element of the pathogenetic cascade. Nuclear arrangement of chromosomes cannot be adequately addressed without I-FISH-based techniques. These studies are valuable for understanding genetic processes occurring in the interphase nucleus. Moreover, it is highly likely that exogenous influencing of chromosomal arrangement in interphase nuclei is a therapeutic opportunity for diseases associated with chromosomal imbalances, susceptibility to chromosome/genome instability,

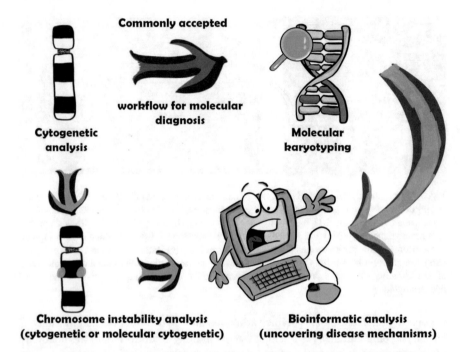

Fig. 7.8 Schematic depiction of the algorithm for investigating the molecular and cellular mechanisms of diseases mediated by CIN. To succeed, one has to follow green arrows or, in other words, to analyze chromosome instability by karyotyping and FISH (analysis of larger amounts of cells) instead of the commonly accepted workflow including only cytogenetic karyotyping and molecular karyotyping; bioinformatics is mandatory for uncovering disease mechanisms. (Copyright © Vorsanova et al. 2019; an open access article distributed under the conditions of the Creative Commons by Attribution License, which permits unrestricted use, distribution, and reproduction in any medium or format, provided the original work is correctly cited)

altered programmed cell death, and abnormal chromatin remodeling. In total, one can conclude that interphase molecular cytogenetics possesses actual methodology for basic and diagnostic research in genetics/genomics, cellular and molecular biology, and molecular (genome) medicine despite the availability of postgenomic technologies.

Acknowledgments We would like to express our gratitude to Dr. MA Zelenova for help in chapter preparation. Prof. SG Vorsanova, Dr. OS Kurinnaia, and Prof. IY Iourov are partially supported by RFBR and CITMA according to the research project No. 18-515-34005. Prof. IY Iourov's lab is supported by the Government Assignment of the Russian Ministry of Science and Higher Education, Assignment no. AAAA-A19-119040490101-6. Prof. SG Vorsanova's lab is supported by the Government Assignment of the Russian Ministry of Health, Assignment no. AAAA-A18-118051590122-7.

References

Andriani GA, Maggi E, Piqué D et al (2019) A direct comparison of interphase FISH versus low-coverage single cell sequencing to detect aneuploidy reveals respective strengths and weaknesses. Sci Rep 9(1):10508

Arendt T, Mosch B, Morawski M (2009) Neuronal aneuploidy in health and disease: a cytomic approach to understand the molecular individuality of neurons. Int J Mol Sci 10(4):1609–1627

Bakker B, van den Bos H, Lansdorp PM et al (2015) How to count chromosomes in a cell: an overview of current and novel technologies. BioEssays 37(5):570–577

Baumgartner A, Ferlatte Hartshorne C, Polyzos A et al (2018) Full karyotype interphase cell analysis. J Histochem Cytochem 66(8):595–606

Bint SM, Davies AF, Ogilvie CM (2013) Multicolor banding remains an important adjunct to array CGH and conventional karyotyping. Mol Cytogenet 6(1):55

Cheng L, Zhang S, Wang L et al (2017) Fluorescence in situ hybridization in surgical pathology: principles and applications. J Pathol Clin Res 3(2):73–99

Chrzanowska NM, Kowalewski J, Lewandowska MA (2020) Use of fluorescence in situ hybridization (FISH) in diagnosis and tailored therapies in solid tumors. Molecules 25(8):1864

Cui C, Shu W, Li P (2016) Fluorescence in situ hybridization: cell-based genetic diagnostic and research applications. Front Cell Dev Biol 4:89

Durm M, Haar F-M, Hausmann M et al (1997) Optimized Fast-FISH with a-satellite probes: acceleration by microwave activation. Braz J Med Biol Res 30(1):15–22

Feuk L, Marshall CR, Wintle RF et al (2006) Structural variants: changing the landscape of chromosomes and design of disease studies. Hum Mol Genet 15(1):R57–R66

Frickmann H, Zautner AE, Moter A et al (2017) Fluorescence in situ hybridization (FISH) in the microbiological diagnostic routine laboratory: a review. Crit Rev Microbiol 43(3):263–293

Gawad C, Koh W, Quake SR (2016) Single-cell genome sequencing: current state of the science. Nat Rev Genet 17(3):175–188

Gersen SL, Keagle MB (2005) The principles of clinical cytogenetics, 2nd edn. Humana Press, Totowa

Gupta P, Balasubramaniam N, Chang HY et al (2020) A single-neuron: current trends and future prospects. Cell 9:1528

Heng HH (2020) New data collection priority: focusing on genome-based bioinformation. Res Result Biomed 6(1):5–8

Hovhannisyan GG (2010) Fluorescence in situ hybridization in combination with the comet assay and micronucleus test in genetic toxicology. Mol Cytogenet 3:17

Hu Q, Maurais EG, Ly P (2020) Cellular and genomic approaches for exploring structural chromosomal rearrangements. Chromosom Res 28(1):19–30

Iourov IY (2012) To see an interphase chromosome or: how a disease can be associated with specific nuclear genome organization. BioDiscovery 4:e8932

Iourov IY (2016) Post genomics: towards a personalized approach to chromosome abnormalities. J Down Syndr Chromosom Abnorm 2(1):2:e104

Iourov IY (2017) Quantitative fluorescence in situ hybridization (QFISH). Methods Mol Biol 1541:143–149

Iourov IY (2019a) Cytogenomic bioinformatics: practical issues. Curr Bioinformatics 14(5):372–373

Iourov IY (2019b) Cytopostgenomics: what is it and how does it work? Curr Genomics 20(2):77–78

Iourov IY, Soloviev IV, Vorsanova SG et al (2005) An approach for quantitative assessment of fluorescence in situ hybridization (FISH) signals for applied human molecular cytogenetics. J Histochem Cytochem 53:401–408

Iourov IY, Liehr T, Vorsanova SG et al (2006a) Visualization of interphase chromosomes in postmitotic cells of the human brain by multicolour banding (MCB). Chromosom Res 14(3):223–229

Iourov IY, Vorsanova SG, Pellestor F et al (2006b) Brain tissue preparations for chromosomal PRINS labeling. Methods Mol Biol 334:123–132

Iourov IY, Vorsanova SG, Yurov YB (2006c) Chromosomal variation in mammalian neuronal cells: known facts and attractive hypotheses. Int Rev Cytol 249:143–191

Iourov IY, Vorsanova SG, Yurov YB (2006d) Intercellular genomic (chromosomal) variations resulting in somatic mosaicism: mechanisms and consequences. Curr Genomics 7:435–446

Iourov IY, Liehr T, Vorsanova SG et al (2007) Interphase chromosome-specific multicolor banding (ICS-MCB): a new tool for analysis of interphase chromosomes in their integrity. Biomol Eng 24(4):415–417

Iourov IY, Vorsanova SG, Yurov YB (2008a) Chromosomal mosaicism goes global. Mol Cytogenet 1:26

Iourov IY, Vorsanova SG, Yurov YB (2008b) Molecular cytogenetics and cytogenomics of brain diseases. Curr Genomics 9(7):452–465

Iourov IY, Vorsanova SG, Yurov YB (2008c) Recent patents on molecular cytogenetics. Recent Pat DNA Gene Seq 2(1):6–15

Iourov IY, Vorsanova SG, Liehr T et al (2009a) Increased chromosome instability dramatically disrupts neural genome integrity and mediates cerebellar degeneration in the ataxia-telangiectasia brain. Hum Mol Genet 18(14):2656–2669

Iourov IY, Vorsanova SG, Liehr T et al (2009b) Aneuploidy in the normal, Alzheimer's disease and ataxia-telangiectasia brain: differential expression and pathological meaning. Neurobiol Dis 34(2):212–220

Iourov IY, Vorsanova SG, Solov'ev IV et al (2010a) Methods of molecular cytogenetics for studying interphase chromosome in human brain cells. Russ J Genet 46(9):1039–1041

Iourov IY, Vorsanova SG, Yurov YB (2010b) Somatic genome variations in health and disease. Curr Genomics 11:387–396

Iourov IY, Vorsanova SG, Yurov YB (2011) Genomic landscape of the Alzheimer's disease brain: chromosome instability – aneuploidy, but not tetraploidy – mediates neurodegeneration. Neurodegener Dis 8:35–37

Iourov IY, Vorsanova SG, Yurov YB (2012) Single cell genomics of the brain: focus on neuronal diversity and neuropsychiatric diseases. Curr Genomics 13(6):477–488

Iourov IY, Vorsanova SG, Yurov YB (2013a) Somatic cell genomics of brain disorders: a new opportunity to clarify genetic-environmental interactions. Cytogenet Genome Res 139(3):181–188

Iourov IY, Vorsanova SG, Voinova VY et al (2013b) Xq28 (MECP2) microdeletions are common in mutation-negative females with Rett syndrome and cause mild subtypes of the disease. Mol Cytogenet 6(1):53

Iourov IY, Vorsanova SG, Liehr T et al (2014a) Mosaike im Gehirn des Menschen. Diagnostische Relevanz in der Zukunft? Med Genet 26(3):342–345

Iourov IY, Vorsanova SG, Yurov YB (2014b) In silico molecular cytogenetics: a bioinformatic approach to prioritization of candidate genes and copy number variations for basic and clinical genome research. Mol Cytogenet 7(1):98

Iourov IY, Vorsanova SG, Demidova IA et al (2015a) 5p13.3p13.2 duplication associated with developmental delay, congenital malformations and chromosome instability manifested as low-level aneuploidy. Springerplus 4(1):616

Iourov IY, Vorsanova SG, Voinova VY et al (2015b) 3p22.1p21.31 microdeletion identifies CCK as Asperger syndrome candidate gene and shows the way for therapeutic strategies in chromosome imbalances. Mol Cytogenet 8:82

Iourov IY, Vorsanova SG, Yurov YB (2017) Interphase FISH for detection of chromosomal mosaicism. In: Liehr T (ed) Fluorescence in situ hybridization (FISH) – application guide (springer protocols handbooks), 2nd edn. Springer, Berlin/Heidelberg, pp 361–372

Iourov IY, Liehr T, Vorsanova SG et al (2019a) The applicability of interphase chromosome-specific multicolor banding (ICS-MCB) for studying neurodevelopmental and neurodegenerative disorders. Res Result Biomed 5(3):4–9

Iourov IY, Vorsanova SG, Yurov YB (2019b) Pathway-based classification of genetic diseases. Mol Cytogenet 12(4)

Iourov IY, Vorsanova SG, Yurov YB (2019c) The variome concept: focus on CNVariome. Mol Cytogenet 12:52

Iourov IY, Vorsanova SG, Yurov YB et al (2019d) Ontogenetic and pathogenetic views on somatic chromosomal mosaicism. Genes (Basel) 10(5):E379

Iurov II, Vorsanova SG, Solov'ev IV et al (2011) Original molecular cytogenetic approach to determining spontaneous chromosomal mutations in the interphase cells to evaluate the mutagenic activity of environmental factors. Gig Sanit 5:90–94

Knouse KA, Wu J, Whittaker CA et al (2014) Single cell sequencing reveals low levels of aneuploidy across mammalian tissues. Proc Natl Acad Sci U S A 111:13409–13414

Liehr T (2017) Fluorescence *in situ* hybridization (FISH) – application Guide. Springer, Berlin/Heidelberg

Liehr T, Al-Rikabi A (2019) Mosaicism: reason for normal phenotypes in carriers of small supernumerary marker chromosomes with known adverse outcome. A systematic review. Front Genet 10:1131

Liehr T, Heller A, Starke H et al (2002) Microdissection based high resolution multicolor banding for all 24 human chromosomes. Int J Mol Med 9(4):335–339

Liehr T, Starke H, Weise A et al (2004) Multicolor FISH probe sets and their applications. Histol Histopathol 19(1):229–237

Liehr T, Othman MA, Rittscher K et al (2015) The current state of molecular cytogenetics in cancer diagnosis. Expert Rev Mol Diagn 15(4):517–526

Martin CL, Warburton D (2015) Detection of chromosomal aberrations in clinical practice: from karyotype to genome sequence. Annu Rev Genomics Hum Genet 16:309–326

Moffitt JR, Hao J, Bambah-Mukku D et al (2016) High-performance multiplexed fluorescence *in situ* hybridization in culture and tissue with matrix imprinting and clearing. Proc Natl Acad Sci U S A 113(50):14456–14461

Pinkel D, Straume T, Gray JW (1986) Cytogenetic analysis using quantitative, high-sensitivity, fluorescence hybridization. Proc Natl Acad Sci U S A 83(9):2934–2938

Riegel M (2014) Human molecular cytogenetics: from cells to nucleotides. Genet Mol Biol 37(1):194–209

Rouquette J, Cremer C, Cremer T et al (2010) Functional nuclear architecture studied by microscopy: present and future. Int Rev Cell Mol Biol 282:1–90

Russo R, Sessa AM, Fumo R et al (2016) Chromosomal anomalies in early spontaneous abortions: interphase FISH analysis on 855 FFPE first trimester abortions. Prenat Diagn 36(2):186–191

Savic S, Bubendorf L (2016) Common fluorescence *in situ* hybridization applications in cytology. Arch Pathol Lab Med 140(12):1323–1330

Soloviev IV, Yurov YB, Vorsanova SG et al (1994) Microwave activation of fluorescence *in situ* hybridization: a novel method for rapid chromosome detection and analysis. Focus 16(4):115–116

Soloviev IV, Yurov YB, Vorsanova SG et al (1995) Prenatal diagnosis of trisomy 21 using interphase fluorescence *in situ* hybridization of post-replicated cells with site-specific cosmid and cosmid contig probes. Prenat Diagn 15:237–248

Soloviev IV, Yurov YB, Vorsanova SG et al (1998) Fluorescent *in situ* hybridization analysis of α-satellite DNA in cosmid libraries specific for human chromosomes 13, 21 and 22. Rus J Genet 34:1247–1255

van der Ploeg M (2000) Cytochemical nucleic acid research during the twentieth century. Eur J Histochem 44(1):7–42

Viotti M (2020) Preimplantation genetic testing for chromosomal abnormalities: aneuploidy, mosaicism, and structural rearrangements. Genes 11:602

Vorsanova SG, Yurov YB, Alexandrov IA et al (1986) 18p- syndrome: an unusual case and diagnosis by *in situ* hybridization with chromosome 18-specific alphoid DNA sequence. Hum Genet 72:185–187

Vorsanova SG, Yurov YB, Kolotii AD et al (2001a) FISH analysis of replication and transcription of chromosome X loci: new approach for genetic analysis of Rett syndrome. Brain and Development 23:S191–S195

Vorsanova SG, Yurov YB, Ulas VY et al (2001b) Cytogenetic and molecular-cytogenetic studies of Rett syndrome (RTT): a retrospective analysis of a Russian cohort of RTT patients (the investigation of 57 girls and three boys). Brain and Development 23:S196–S201

Vorsanova SG, Iourov IY, Beresheva AK et al (2005a) Non-disjunction of chromosome 21, alphoid DNA variation, and sociogenetic features of Down syndrome. Tsitol Genet 39(6):30–36

Vorsanova SG, Kolotii AD, Iourov IY et al (2005b) Evidence for high frequency of chromosomal mosaicism in spontaneous abortions revealed by interphase FISH analysis. J Histochem Cytochem 53(3):375–380

Vorsanova SG, Yurov IY, Demidova IA et al (2007) Variability in the heterochromatin regions of the chromosomes and chromosomal anomalies in children with autism: identification of genetic markers of autistic spectrum disorders. Neurosci Behav Physiol 37(6):553–558

Vorsanova SG, Iourov IY, Kolotii AD et al (2010a) Chromosomal mosaicism in spontaneous abortions: analysis of 650 cases. Rus J Genet 46:1197–1200

Vorsanova SG, Voinova VY, Yurov IY et al (2010b) Cytogenetic, molecular-cytogenetic, and clinical-genealogical studies of the mothers of children with autism: a search for familial genetic markers for autistic disorders. Neurosci Behav Physiol 40(7):745–756

Vorsanova SG, Yurov YB, Iourov IY (2010c) Human interphase chromosomes: a review of available molecular cytogenetic technologies. Mol Cytogenet 3:1

Vorsanova SG, Yurov YB, Soloviev IV et al (2010d) Molecular cytogenetic diagnosis and somatic genome variations. Curr Genomics 11(6):440–446

Vorsanova SG, Yurov YB, Iourov IY (2013) Technological solutions in human interphase cytogenetics. In: Yurov YB, Vorsanova SG, Iourov IY (eds) Human interphase chromosomes (biomedical aspects). Springer, New York/Heidelberg/Dordrecht/London, pp 179–203

Vorsanova SG, Yurov YB, Soloviev IV et al (2019) FISH-based analysis of mosaic aneuploidy and chromosome instability for investigating molecular and cellular mechanisms of disease. OBM Genetics 3(1):9

Wan TS (2017) Cancer cytogenetics. Springer, New York

Weise A, Liehr T, Claussen U et al (2005) Increased efficiency of fluorescence in situ hybridization (FISH) using the microwave. J Histochem Cytochem 53(10):1301–1303

Weise A, Mrasek K, Pentzold C et al (2019) Chromosomes in the DNA era: perspectives in diagnostics and research. Med Genet 31(1):8–19

Ye CJ, Stilgenbauer L, Moy A et al (2019) What is karyotype coding and why is genomic topology important for cancer and evolution? Front Genet 10:1082

Yurov YB, Soloviev IV, Vorsanova SG et al (1996) High resolution multicolor fluorescence in situ hybridization using cyanine and fluorescein dyes: rapid chromosome identification by directly fluorescently labeled alphoid DNA probes. Hum Genet 97(3):390–398

Yurov YB, Vostrikov VM, Vorsanova SG et al (2001) Multicolor fluorescent in situ hybridization on post-mortem brain in schizophrenia as an approach for identification of low-level chromosomal aneuploidy in neuropsychiatric diseases. Brain and Development 23(1):S186–S190

Yurov YB, Iourov IY, Monakhov VV et al (2005) The variation of aneuploidy frequency in the developing and adult human brain revealed by an interphase FISH study. J Histochem Cytochem 53(3):385–390

Yurov YB, Iourov IY, Vorsanova SG et al (2007a) Aneuploidy and confined chromosomal mosaicism in the developing human brain. PLoS One 2(6):e558

Yurov YB, Vorsanova SG, Iourov IY et al (2007b) Unexplained autism is frequently associated with low-level mosaic aneuploidy. J Med Genet 44(8):521–525

Yurov YB, Iourov IY, Vorsanova SG et al (2008) The schizophrenia brain exhibits low-level aneuploidy involving chromosome 1. Schizophr Res 98:139–147

Yurov YB, Vorsanova SG, Iourov IY (2009) GIN'n'CIN hypothesis of brain aging: deciphering the role of somatic genetic instabilities and neural aneuploidy during ontogeny. Mol Cytogenet 2:23

Yurov YB, Vorsanova SG, Iourov IY (2010a) Ontogenetic variation of the human genome. Curr Genomics 11(6):420–425

Yurov YB, Vorsanova SG, Solov'ev IV et al (2010b) Instability of chromosomes in human nerve cells (normal and with neuromental diseases). Russ J Genet 46(10):1194–1196

Yurov YB, Vorsanova SG, Iourov IY (2013) Human interphase chromosomes – biomedical aspects. Springer, New York/Heidelberg/Dordrecht/London

Yurov YB, Vorsanova SG, Liehr T et al (2014) X chromosome aneuploidy in the Alzheimer's disease brain. Mol Cytogenet 7(1):20

Yurov YB, Vorsanova SG, Soloviev IV et al (2017) FISH-based assays for detecting genomic (chromosomal) mosaicism in human brain cells. NeuroMethods 131:27–41

Yurov YB, Vorsanova SG, Demidova IA et al (2018a) Mosaic brain aneuploidy in mental illnesses: an association of low-level post-zygotic aneuploidy with schizophrenia and comorbid psychiatric disorders. Curr Genomics 19(3):163–172

Yurov YB, Vorsanova SG, Iourov IY (2018b) Human molecular neurocytogenetics. Curr Genet Med Rep 6(4):155–164

Yurov YB, Vorsanova SG, Iourov IY (2019a) Chromosome instability in the neurodegenerating brain. Front Genet 10:892

Yurov YB, Vorsanova SG, Iourov IY (2019b) FISHing for unstable cellular genomes in the human brain. OBM Genetics 3(2):11

Zeng H, Weier JF, Wang M et al (2012) Bioinformatic tools identify chromosome-specific DNA probes and facilitate risk assessment by detecting aneusomies in extra-embryonic tissues. Curr Genomics 13(6):438–445

Zhang C, Cerveira E, Rens W et al (2018) Multicolor fluorescence *in situ* hybridization (FISH) approaches for simultaneous analysis of the entire human genome. Curr Protoc Hum Genet 99(1):e70

Chapter 8
Chromosome Architecture Studied by High-Resolution FISH Banding in Three-Dimensionally Preserved Human Interphase Nuclei

Thomas Liehr

Abstract The impact of chromosome architecture in the formation of chromosome aberrations is a meanwhile well-established finding of interphase-directed molecular cytogenetic studies. Up to recent years, biomedical research of interphase chromosomes in their integrity was hindered by technical limitations. The introduction of three-dimensional suspension-based fluorescence in situ hybridization (S-FISH) in combination with microdissection-based engineered DNA probes and fluorescence multicolor chromosome banding (MCB) allowed studying interphase chromosome organization, numbers, and rearrangements in different kind of cells. Such studies already provided comprehensive information on the interphase architecture of normal human sperm, as well as first insights into the influence of chromosomal rearrangements on the 3D structure of the sperm nuclei. Also, the influence of additional chromosomal fragments present in a nucleus was successfully visualized by S-FISH. Finally, S-FISH supported the idea that disease-specific chromosomal translocations could be due to tissue specific genomic organization. Overall, S-FISH combined with MCB but also other DNA probes is a tool with high potential to resolve the influence of chromosomal imbalances and/or rearrangements on the interphase architecture, the latter being possibly a part of the epigenetic cell regulation, also being denominated as chromosomics.

Introduction

In the interphase nucleus, chromosomes are located in specific regions, which are called "chromosome territories" (Cremer and Cremer 2001; Williams and Fisher 2003; Branco and Pombo 2006). Own multicolor banding (MCB)-based studies

T. Liehr (✉)
Jena University Hospital, Friedrich Schiller University, Institute of Human Genetics, Jena, Germany
e-mail: Thomas.Liehr@med.uni-jena.de

© Springer Nature Switzerland AG 2020
I. Iourov et al. (eds.), *Human Interphase Chromosomes*,
https://doi.org/10.1007/978-3-030-62532-0_8

revealed that the chromosome shape itself is not lost in the interphase nucleus, and one can even identify "interphase chromosomes" instead of only chromosome territory, even irrespective of the cell cycle phase (Weise et al. 2002; Lemke et al. 2002).

Both chromosome size and gene density are discussed to have an important impact on the nuclear position of chromosomes. Small chromosomes preferentially locate close to the center of the nucleus, while large chromosomes can be found in the nuclear periphery (Sun et al. 2000; Bolzer et al. 2005). On the other hand, Croft et al. (1999) demonstrated a gene density-correlated radial arrangement of chromosomes in nuclei. Mainly gene-dense and early replicating chromatin can be found in the central part of the nucleus, while gene-poor and later replicating chromatin is located in nuclear periphery (Croft et al. 1999). Interestingly, this nuclear topological arrangement is conserved during primate evolution (Manvelyan et al. 2008a).

Here, we summarize the yet published applications of suspension-based fluorescence in situ hybridization (S-FISH) combined with FISH banding (Liehr et al. 2002, 2006), particularly the yet most used approach array-proven MCB (Weise et al. 2008). Besides, also other protocols were suggested for FISH studies in 3D-preserved nuclei (e.g., Walter et al. 2006). Also, recent studies showed that inter- and metaphase chromosomes preserve a genome-wide haploid order (Weise et al. 2016) and that this order is completely changed in senescent cells (Roediger et al. 2014). All these studies provide to the more and more emerging field of chromosomics, as predicted in 2005 by Prof. Uwe Claussen (Claussen 2005).

S-FISH, the Method

Performing of a FISH experiment on human meta- and interphase cells after air-drying method is a well-established approach; it is routinely done as one- to multicolor-FISH test (Liehr et al. 2004a). However, the air-drying procedure of chromosome preparation, leading to well-spread metaphases under appropriate conditions, leads at the same time to flattening of the originally spherical interphase nuclei. Thus, interphase architecture is hard to be studied reliably on such kind of preparation (Hunstig et al. 2009), even though some basic insights can also be gained using such material for FISH banding (Weise et al. 2002; Lemke et al. 2002).

Still, there is an easy way to do studies in three-dimensionally (3D) preserved interphase nuclei obtained from routinely prepared cytogenetic preparations stored in Carnoy's fixative. One just needs to do the whole FISH procedure in cell suspension, and as a final step, the nuclei are placed on a polished concave slide before evaluation, immobilized in agarose. This approach for 3D-FISH analyses on totally spherical interphase nuclei, called suspension-based fluorescence in situ hybridization (S-FISH), was published first in 2002 (Steinhaeuser et al. 2002) and further developed and slightly modified later (Manvelyan et al. 2008a; Hunstig et al. 2009). Its principle is shown in Fig. 8.1.

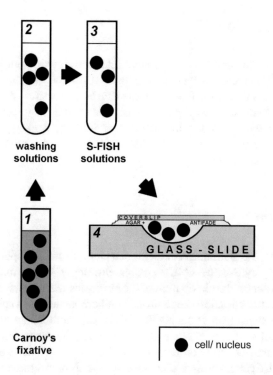

Fig. 8.1 Schematic drawing of the suspension-based fluorescence in situ hybridization (S-FISH) procedure. Overall, S-FISH avoids this flattening and artificial swelling of the interphase nuclei, and the whole experiment is performed in suspension. A certain loss of cells during the washing steps is normal, shown here by the reduction of cells/nuclei from step 1 to step 4. In principle, Carnoy's fixative is replaced subsequently by solutions necessary for a FISH, and washing steps are included. Finally, the cells/nuclei are immobilized and counterstained in an agarose (AGAR) on a glass slide under a coverslip. The details of the protocol are described in Hunstig et al. (2009)

S-FISH: Which DNA Probes May Be Applied?

For S-FISH, all available chromosome or chromosome region-specific DNA are principally suited. However, for application in S-FISH, at least double amount of the probe is necessary than for "normal" FISH experiments (Hunstig et al. 2009). To resolve the chromosome structure as a whole, single chromosome-directed FISH banding based on partial chromosome painting probes like in MCB is suited best (Weise et al. 2008). Besides, centromeric and/or locus-specific probes can be used as well for special questions (e.g., Manvelyan et al. 2009; Hunstig et al. 2009).

Applications of S-FISH

Besides some studies done in comparative interphase cytogenetics of human and whitehanded gibbon and gorilla (Manvelyan et al. 2008a), S-FISH combined with MCB is mainly applied in the field of biomedical basic research of the human interphase nucleus. Here, still many questions are open and unanswered, mainly due to lack of suited methods, before introduction of S-FISH. Besides, more and more studies in other animals/species provide insights into the nuclear architecture (Karamysheva et al. 2017).

Human Sperm

For the first time, the distribution of all human chromosomes in sperm was resolved comprehensively by S-FISH-/MCB studies. Strikingly, for the majority of the 24 human chromosomes, the distribution of the territories was alike as in lymphocytes; only the acrocentric chromosomes showed another location as in sperm, no nucleolus is formed (Manvelyan et al. 2008b). Thus, this nonrandom positioning must have a biological meaning. In other words, each chromosome needs to have a special position in the nucleus in order that the cell can work properly. Sperm are translationally inactive cells; however, they need to have chromosomes at the right places as soon as a sperm enters an oocyte and needs to become active again.

The study of Manvelyan et al. (2008b) showed a direct correlation of chromosome positions and their sizes, apart from chromosomes 1, 2, 6, 14, 18, 20, 21, and Y, i.e., large chromosomes were in the periphery, small in the center. Exactly those eight chromosomes not fitting in the correlation before perfectly aligned with gene density theory, i.e., gene-dense chromosomes were in the nuclear center, and gene-poor in the periphery.

There are also already other one studies in sperm of male with a chromosomal aberration (Bhatt et al. 2009; Karamysheva et al. 2015). Three males with paracentric inversion were studied, and no gross changes in the interphase positioning of the affected chromosomes were found. Here for sure, more studies on the influence of inborn rearrangements on the nuclear architecture of sperm, but also other in tissues, are necessary.

Different Tissues with Additional Chromosomal Fragments

Additional chromosomal material present in the cell is suspected to alter or at least influence the chromosomal architecture. Besides complete trisomies as inborn or acquired aberrations, there is the possibility of partial trisomies induced either by derivative chromosomes or by the presence of a small supernumerary marker chromosome (sSMC). The latter condition may be seen in 0.043% of newborn infants, 0.077% of prenatal cases, 0.433% of mentally retarded patients, and 0.171% of subfertile people (Liehr and Weise 2007). sSMC are defined as structurally abnormal chromosomes that cannot be identified or characterized unambiguously by conventional banding cytogenetics alone and are generally equal in size or smaller than a chromosome 20 of the same metaphase spread. sSMC are mostly detected unexpectedly in routine cytogenetics (Liehr et al. 2004b). Also, they are not easy to correlate with a specific clinical outcome as besides induction of genomic imbalance, mosaicism and other most often epigenetic factors can influence the phenotype of an sSMC carrier: Uniparental disomy, heterochromatization, and even their influence on the interphase architecture may play a role here. Also, a pilot study revealed some potential influence of sSMC presence on nuclear architecture recently (Karamysheva et al. 2015).

In a recent study (Klein et al. 2012), S-FISH revealed that an extra piece of DNA like an sSMC leads to gross rearrangements within the interphase nucleus, mainly concerning the sSMCs' normal sister chromosomes. Primarily, the position of the sSMC is influenced by and/or influencing the position of the homologous chromosomes. sSMC and one sister chromosome tend to colocalize; this seems to be driven mainly by the amount of euchromatin present in the sSMC. Also, the sSMC seems to take over the position of one normal sister chromosome. Thus, the remainder sister chromosome is displaced toward another location within the nucleus. These observations were made in B and T lymphocytes and/or skin fibroblasts.

Different Female Tissues and the Position of the X Chromosome

S-FISH/MCB studies in buccal mucosa, B and T lymphocytes, and skin fibroblasts for the positioning of normal and derivative X chromosomes in female cells also may lead to interesting, yet impossible insights into the nuclear architecture. Preliminary yet unpublished results (Fig. 8.2) firstly confirmed that active and inactive X chromosomes are located in the cell periphery and that the inactive X chromosome colocalizes to big parts, even though not perfectly, with the Barr body. Interestingly, a dicentric X chromosome, leading to an almost complete trisomy X, altered the positioning of the two X chromosomes to each other, inducing a larger distance between both normal and derivative X chromosome compared to the normal cells. Thus, new insights may be obtained also by studying well-known phenomenon like X inactivation by the S-FISH approach.

Leukemia and the Positions of Chromosomes 8 and 21

Nonrandom positioning of chromosomes in interphase nuclei is known to be of importance for genomic stability and formation of chromosome aberrations. So tissue specificity of chromosomal translocations could be due to tissue-specific

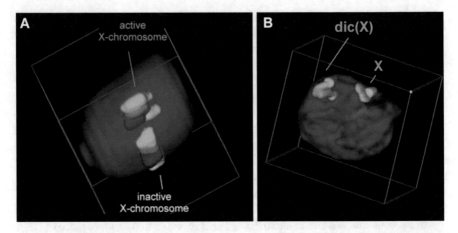

Fig. 8.2 S-FISH results after application of X chromosome-specific DNA probe sets. (**a**) Active and inactive X chromosomes in a lymphocyte nucleus of a normal female labeled with an MCB-X probe set. (**b**) A normal (X) and derivative X chromosome (dic(X)) labeled with partial chromosome paints for Xp (green) and Xq (yellow) visualized in the fibroblast cell line GM15859 (Coriell). The female carrier had a constitutional karyotype 46,X,dic(X)(pter->q28::q28->pter)

genome organization (Meaburn et al. 2007; Brianna Caddle et al. 2007), and a positive correlation between spatial proximity of chromosomes/genes in interphase nuclei and translocation frequencies was shown (Bickmore and Teague 2002; Roix et al. 2003; Branco and Pombo 2006; Meaburn et al. 2007; Brianna Caddle et al. 2007; Grasser et al. 2008).

Manvelyan et al. (2008a, b) provided evidence that there might be an effect of specific chromosome positioning in myeloid bone marrow cells, i.e., a colocalization of chromosomes 8 and 21 could promote a translocation providing selective advantage of t(8;21) cells in AML-M2. Additional S-FISH studies confirmed that this is specifically true for AML patients having a trisomy 8 (Othman et al. 2012). Overall, studies to enlighten the nuclear position of tumor-related oncogenes, which are known to be activated by specific translocations are promising targets of future S-FISH-studies, as supported by recent comparable findings in thyroid cancer (Gandhi et al. 2009).

S-FISH, Conclusions, and Perspectives

Overall, the combination of S-FISH and MCB for a three-dimensional analysis of chromosome position in interphase nucleus is a powerful tool, which can be accompanied by the use of locus-specific probes. The topological organization in interphase nucleus is nonrandom, and it becomes more and more obvious that there is a physiological reason behind that.

The already done and above summarized S-FISH studies in human show the potential of this approach for (i) genome-wide analysis of interphase architecture in yet not studied tissues (like done for sperm (Manvelyan et al. 2008b)), (ii) studies on architectural changes in nuclei with additional chromosomes or chromosomal material (like done for sSMC (Klein et al. 2012; Karamysheva et al. 2015) or the X chromosome), and (iii) analysis for the susceptibility of specific parts of the human genome for rearrangements due to colocalization (like done for the t(8;21) in AML (Manvelyan et al. 2009; Othman et al. 2012)). For sure, additional biomedical research aspect of interphase chromosomes may also be covered using the S-FISH/MCB approach, like recently the proof of interaction between distant chromosomal regions (Maass et al. 2018) and the description of nuclear architecture in hematopoietic stem cells (Grigoryan et al. 2018).

Overall, the approach discussed can be used not only based on human but also, if MCB probes are available for, based on probes from other species as already demonstrated by one example for murine mcb (Ktistaki et al. 2010). In conclusion, big advances in the field of chromosomics can be expected in the future from high-resolution FISH banding (MCB/mcb) in three-dimensionally preserved human interphase nuclei.

References

Bhatt S, Moradkhani K, Mrasek K, Puechberty J, Manvelyan M, Hunstig F, Lefort G, Weise A, Lespinasse J, Sarda P, Liehr T, Hamamah S, Pellestor F (2009) Breakpoint mapping and complete analysis of meiotic segregation patterns in three men heterozygous for paracentric inversions. Eur J Hum Genet 17:44–50

Bickmore WA, Teague P (2002) Influences of chromosome size, gene density and nuclear position on the frequency of constitutional translocations in the human population. Chromosom Res 10:707–715

Bolzer A, Kreth G, Solovei I, Koehler D, Saracoglu K, Fauth C, Muller S, Eils R, Cremer C, Speicher MR, Cremer T (2005) Three-dimensional maps of all chromosomes in human male fibroblast nuclei and prometaphase rosettes. PLoS Biol 3:e157

Branco MR, Pombo A (2006) Intermingling of chromosome territories in interphase suggests role in translocations and transcription-dependent associations. PLoS Biol 4:e138

Brianna Caddle L, Grant JL, Szatkiewicz J, van Hase J, Shirley BJ, Bewersdorf J, Cremer C, Arneodo A, Khalil A, Mills KD (2007) Chromosome neighborhood composition determines translocation outcomes after exposure to high-dose radiation in primary cells. Chromosom Res 15:1061–1073

Claussen U (2005) Chromosomics. Cytogenet Genome Res 111:101–106

Cremer T, Cremer C (2001) Chromosome territories, nuclear architecture and gene regulation in mammalian cells. Nat Rev Genet 2:292–301

Croft JA, Bridger JM, Boyle S, Perry P, Teague P, Bickmore WA (1999) Differences in the localization and morphology of chromosomes in the human nucleus. J Cell Biol 45:1119–1131

Gandhi MS, Stringer JR, Nikiforova M, Medvedovic M, Nikiforov YE (2009) Gene position within chromosome territories correlates with their involvement in distinct rearrangement types in thyroid cancer cells. Genes Chromosom Cancer 48:222–228

Grasser F, Neusser M, Fiegler H, Thormeyer T, Cremer M, Carter NP, Cremer T, Miller S (2008) Replication-timing correlated spatial chromatin arrangements in cancer and in primate interphase nuclei. J Cell Sci 121:1876–1886

Grigoryan A, Guidi N, Senger K, Liehr T, Soller K, Marka G, Vollmer A, Markaki Y, Leonhardt H, Buske C, Lipka DB, Plass C, Zheng Y, Mulaw MA, Geiger H, Florian MC (2018) LaminA/C regulates epigenetic and chromatin architecture changes upon aging of hematopoietic stem cells. Genome Biol 19:189

Hunstig F, Manvelyan M, Bhatt S, Steinhaeuser U, Liehr T (2009) Three-dimensional interphase analysis enabled by suspension FISH. In: Liehr T (ed) Fluorescence in situ hybridization (FISH) – application guide, 1st edn. Springer, Berlin

Karamysheva T, Kosyakova N, Guediche N, Liehr T (2015) Small supernumerary marker chromosomes and the nuclear architecture of sperm – a study in a fertile and an infertile brother. Syst Biol Reprod Med 61:32–36

Karamysheva TV, Torgasheva AA, Yefremov YR, Bogomolov AG, Liehr T, Borodin PM, Rubtsov NB (2017) Spatial organization of fibroblast and spermatocyte nuclei with different B-chromosome content in Korean field mouse, Apodemus peninsulae (Rodentia, Muridae). Genome 60:815–824

Klein E, Manvelyan M, Simonyan I, Hamid AB, Santos Guilherme R, Liehr T, Karamysheva T (2012) Centromeric association of small supernumerary marker chromosomes with their sister-chromosomes detected by three dimensional molecular cytogenetics. Mol Cytogenet 5:15

Ktistaki E, Garefalaki A, Williams A, Andrews SR, Bell DM, Foster KE, Spilianakis CG, Flavell RA, Kosyakova N, Trifonov V, Liehr T, Kioussis D (2010) CD8 locus nuclear dynamics during thymocyte development. J Immunol 184:5686–5695

Lemke J, Claussen J, Michel S, Chudoba I, Mühlig P, Westermann M, Sperling K, Rubtsov N, Grummt UW, Ullmann P, Kromeyer-Hauschild K, Liehr T, Claussen U (2002) The DNA-based structure of human chromosome 5 in interphase. Am J Hum Genet 71:1051–1059

Liehr T, Weise A (2007) Frequency of small supernumerary marker chromosomes in prenatal, newborn, developmentally retarded and infertility diagnostics. Int J Mol Med 19:719–731

Liehr T, Heller A, Starke H, Claussen U (2002) FISH banding methods: applications in research and diagnostics. Expert Rev Mol Diagn 2:217–225

Liehr T, Starke H, Weise A, Lehrer H, Claussen U (2004a) Multicolor FISH probe sets and their applications. Histol Histopathol 19:229–237

Liehr T, Claussen U, Starke H (2004b) Small supernumerary marker chromosomes (sSMC) in humans. Cytogenet Genome Res 107:55–67

Liehr T, Starke H, Heller A, Kosyakova N, Mrasek K, Gross M, Karst C, Steinhaeuser U, Hunstig F, Fickelscher I, Kuechler A, Trifonov V, Romanenko SA, Weise A (2006) Multicolor fluorescence in situ hybridization (FISH) applied to FISH-banding. Cytogenet Genome Res 114:240–244

Maass PG, Weise A, Rittscher K, Lichtenwald J, Barutcu AR, Liehr T, Aydin A, Wefeld-Neuenfeld Y, Pölsler L, Tinschert S, Rinn JL, Luft FC, Bähring S (2018) Reorganization of interchromosomal interactions in the 2q37-deletion syndrome. EMBO J 37:e96257

Manvelyan M, Hunstig F, Mrasek K, Bhatt S, Pellestor F, Weise A, Liehr T (2008a) Position of chromosomes 18, 19, 21 and 22 in 3D-preserved interphase nuclei of human and gorilla and white hand gibbon. Mol Cytogenet 1:9

Manvelyan M, Hunstig F, Bhatt S, Mrasek K, Pellestor F, Weise A, Simonyan I, Aroutiounian R, Liehr T (2008b) Chromosome distribution in human sperm – a 3D multicolor banding-study. Mol Cytogenet 1:25

Manvelyan M, Kempf P, Weise A, Mrasek K, Heller A, Lier A, Höffken K, Fricke HJ, Sayer HG, Liehr T, Mkrtchyan H (2009) Preferred co-localization of chromosome 8 and 21 in myeloid bone marrow cells detected by three dimensional molecular cytogenetics. Int J Mol Med 24:335–341

Meaburn KJ, Misteli T, Soutoglou E (2007) Spatial genome organization in the formation of chromosomal translocations. Semin Cancer Biol 17:80–90

Othman MAK, Lier A, Junker S, Kempf P, Dorka F, Gebhart E, Sheth FJ, Grygalewicz B, Bhatt S, Weise A, Mrasek K, Liehr T, Manvelyan M (2012) Does positioning of chromosomes 8 and 21 in interphase drive t(8;21) in acute myelogenous leukemia? BioDiscovery 4:4

Roediger J, Hessenkemper W, Bartsch S, Manvelyan M, Huettner SS, Liehr T, Esmaeili M, Foller S, Petersen I, Grimm MO, Baniahmad A (2014) Supraphysiological androgen levels induce cellular senescence in human prostate cancer cells through the Src-Akt pathway. Mol Cancer 13:214

Roix JJ, McQueen PG, Munson PJ, Parada LA, Misteli T (2003) Spatial proximity of translocation-prone gene loci in human lymphomas. Nat Genet 34:287–291

Steinhaeuser U, Starke H, Nietzel A, Lindenau J, Ullmann P, Claussen U, Liehr T (2002) Suspension (S)-FISH, a new technique for interphase nuclei. J Histochem Cytochem 50:1697–1698

Sun HB, Shen J, Yokota H (2000) Size-dependent positioning of human chromosomes in interphase nuclei. Biophys J 79:184–190

Walter J, Joffe B, Bolzer A, Albiez H, Benedetti PA, Müller S, Speicher MR, Cremer T, Cremer M, Solovei I (2006) Towards many colors in FISH on 3D-preserved interphase nuclei. Cytogenet Genome Res 114:367–378

Weise A, Starke H, Heller A, Claussen U, Liehr T (2002) Evidence for interphase DNA decondensation transverse to the chromosome axis: a multicolor banding analysis. Int J Mol Med 9:359–361

Weise A, Mrasek K, Fickelscher I, Claussen U, Cheung SW, Cai WW, Liehr T, Kosyakova N (2008) Molecular definition of high-resolution multicolor banding probes: first within the human DNA sequence anchored FISH banding probe set. J Histochem Cytochem 56:487–493

Weise A, Bhatt S, Piaszinski K, Kosyakova N, Fan X, Altendorf-Hofmann A, Tanomtong A, Chaveerach A, de Cioffi MB, de Oliveira E, Walther JU, Liehr T, Chaudhuri JP (2016) Chromosomes in a genome-wise order: evidence for metaphase architecture. Mol Cytogenet 9:36

Williams RE, Fisher AG (2003) Chromosomes, positions please! Nat Cell Biol 5:388–390

Chapter 9
Chromosome-Centric Look at the Genome

Ivan Y. Iourov, Yuri B. Yurov, and Svetlana G. Vorsanova

Abstract Undoubtedly, genome-centric and gene-centric are the words to describe actual concepts in human genetics. In a world of genes and genomes, the lack of required attention to chromosomes is often observed. As a result, chromosome research gradually loses the genetic (genomic) context. Certainly, brilliant insights into chromosome biology obtained by studies dedicated to molecular/cell biology, evolution, biochemistry, biophysics, etc., are fascinating. However, genome research and human (medical) genetics miss the essential link between genes and genomes, which is determined by chromosomal analysis (i.e., cytogenetics, molecular cytogenetics, cytogenomics). This is also the case for diagnostic research, which has recently suffered problems in quality of cytogenetic diagnosis. Ignoring chromosomal and subchromosomal variations creates a blurred vision on genetic etiology of a disease. Data on genes and genomes are useless outside the chromosomal context when intrinsic molecular and cellular pathways are highlighted in health and disease. Without the chromosomal context, genes are virtual elements interacting with each other in an elusive digital universe. Unfortunately, this situation is generally the case for numerous attempts to analyze and interpret genomic data. More dramatically, education programs in genomics and genomic medicine developed for medical/biological students, physicians, or the public generally conceal any information about the chromosome, the physical (biological) storage of genomic data. In our opinion, there is an urgent need for expressing chromosome-centric concepts for filling the "chromosomal gap" in human genetics (genomics) and genomic medicine. To succeed, one has to look at the problem from different perspectives: theoretical, empirical, diagnostic, and educational.

I. Y. Iourov (✉)
Mental Health Research Center, Moscow, Russia

Veltischev Research and Clinical Institute for Pediatrics of the Pirogov Russian National Research Medical University, Ministry of Health of Russian Federation, Moscow, Russia

Medical Genetics Department of Russian Medical Academy of Continuous Postgraduate Education, Moscow, Russia

Y. B. Yurov · S. G. Vorsanova
Mental Health Research Center, Moscow, Russia

Veltischev Research and Clinical Institute for Pediatrics of the Pirogov Russian National Research Medical University, Ministry of Health of Russian Federation, Moscow, Russia

I. Iourov et al. (eds.), *Human Interphase Chromosomes*,
https://doi.org/10.1007/978-3-030-62532-0_9

Introduction: Where Have All the Chromosomes Gone?

More than a century ago, the chromosome theory of heredity was developed (Bridges 1916). Since that time, cytogenetics (the study of chromosomes) has evolved into a huge area of biomedical research. Actually, chromosomal analyses are relevant to almost all fields of bioscience. Basic and diagnostic research in medicine, molecular and cell biology, evolution, and, more specifically, human genetics and genomics benefit from the knowledge generated by cytogenetics, molecular cytogenetics, and cytogenomics (CM2C[1]) (Gersen and Keagle 2005; Page and Holmes 2009; Liehr 2017, 2019). However, CM2C seem to be excluded from the essential scope ("mainstream") of current human genetics, and genome medicine supposed to have preferentially diagnostic value (Liehr 2019). However, chromosomal banding (appearance of metaphase chromosomes) and the analysis in the CM2C context have long been shown to be important for understanding genome organization and behavior (Korenberg and Rykowski 1988; Bickmore and Sumner 1989; Holmquist 1992; Costantini et al. 2007; Kosyakova et al. 2009; Bernardi 2015; Daban 2015). Furthermore, studying chromosomes at different cell cycle stages (interphase) has demonstrated that chromosomal structural and functional organization mediates behavior, stability, and replication of the nuclear genome (Manuelidis 1990; Sadoni et al. 1999; Vorsanova et al. 2010b; Rodriguez and Bjerling 2013; Yurov et al. 2013; Cook and Marenduzzo 2018; Cremer and Cremer 2019; Jerković et al. 2020). Additionally, systems biology analyses of chromosomal variations allow unravelling genetic/genomic pathways to a wide spectrum of diseases (Iourov et al. 2014a, b, 2019b; Vorsanova et al. 2017; Yurov et al. 2017; Zelenova et al. 2019). In summary, on the one hand, CM2C are exciting fields of biomedical discoveries in medical genetics and genomics, whereas on the other hand, these are the foundations of a world parallel to mainstream research in medical genetics/genomics. Because of this dilemma, one may be curious to learn about the place of chromosome research in postmodern bioscience.

Chromosomal banding (cytogenetic analysis) was the technological basis of the majority of studies performed during the first 30–40 years of the history of empirical genetics (i.e., genetic studies using laboratory methods) from late 1950s to 1990s (Liehr 2017, 2019). Probably, the unavailability of more or less adequate alternatives for cytogenetic analysis has led lately to an erroneous impression that studying human chromosomes is unable to give further insights into medical genetics and genomics. The "golden age" of banding (classical) cytogenetics (1960s–1980s) was the era of human genetics when it was established that a disease may be associated with a genetic defect. In other words, banding cytogenetics was the start of medical genetics as an empirical discipline. Currently, a multitude of pathological conditions are associated with chromosomal abnormalities

---

[1]All three terms mean studies of chromosomes but differ with respect to the technological basis. Since true cytogenetic researchers are urged to become engaged in all these areas, we prefer to represent simultaneously these three biomedical fields by a single abbreviation—CM2C.

(variations). As a result, CM2C data are dispersed over many biomedical areas, i.e., cancer research, reproduction, pediatrics, cardiology, neurology, psychiatry, etc. (Gersen and Keagle 2005; Iourov et al. 2006a, 2008b; Liehr 2013; Gonzales et al. 2016). This is also applicable to genomic variations at the subchromosomal level (e.g., copy number variations or CNVs) (Feuk et al. 2006; Iourov et al. 2008b; Liehr 2013; Savory et al. 2020). Thus, chromosomal and subchromosomal variations seem to be a focus of genetic studies dedicated to specific biomedical fields rather than human/medical genetics per se. It is to note that chromosome abnormalities and CNVs are still considered an important focus of diagnostic research and development of sequencing technologies (Ho et al. 2020; Savory et al. 2020). However, massive genomic data acquired through the last decade are generally out of the chromosomal context (Hochstenbach et al. 2019; Liehr 2019; Heng 2020). Consequently, genomes are considered to be loose sets of genes and noncoding DNAs, whereas genes are considered as protein-coding texts "floating" in immaterial space though genes may be interconnected through networks (pathways) by systems biology analysis. Canonical authors occasionally indicate chromosomal localization of a genomic change. Still, the chromosomal aspects of genome variations (i.e., intrinsic chromosomal localization/neighboring, CNVs of mutated genes, implication in a pathway related to chromosome instability, etc.) are ignored. These genomic studies give further insight neither to consequence of a variation in terms of genomic milieu nor to variation's biological basis (Iourov et al. 2019b). Since chromosomes are the physical (biological) storages of genomic data, the falling out of the scope of genomics and genomic medicine leads to a gap in our knowledge about the real (material) cellular genome. As such, numerous genomic studies (basic and diagnostic) usually deal with a "virtual" genome, which is easily manageable for theoretical and public relations purposes, but this "genome" does not correspond to the "real" genome (i.e., karyotype or complete sets of nuclear/chromosomal DNAs) in a cell. CM2C studies are able to reconcile the concepts of "virtual" and "real" genome.

The idea that sequence-based genome analysis cannot be the unique basis of genomic research is not new (Heng et al. 2011). However, the chromosomal basis of genomic variations has been permanently left aside since the introduction of high-resolution (next-generation) sequencing technologies (Iourov et al. 2006b, 2010; Liehr 2019; Heng 2020). As a result, insights into chromosome biology are brought by a wide spectrum of biomedical disciplines different to human genetics and genomics (Iourov 2019b). The latest knowledge on chromosome variations and behavior has been acquired by a myriad of brilliant studies dedicated to cancer (for more details, see Liehr [2017], Christine et al. [2018], Hnisz et al. [2018], Ye et al. [2019], Umbreit et al. [2020]). Data on ontogenetic (ontogenomic) variability at the chromosomal level have been essentially accumulated during molecular cytogenetic analysis of developing and aging human tissues (Yurov et al. 2007, 2009b, 2010, 2014). Structural origins of human chromosomes and basic principles of the behavior have been uncovered by an enormous amount of evolution studies (Page and Holmes 2009; Liehr 2013; Ye et al. 2019). Cellular/nuclear genome behavior (the behavior of real vs. virtual genome) at the supramolecular or chromosomal level is the focus of cell biology (+molecular biology, biophysics, biochemistry) and is rarely addressed in genomics' context (Chevret et al. 2000; Dixon et al. 2016; Nagano et al. 2017; Knoch 2019; Maass et al. 2019; Szczepińska et al. 2019; Cremer et al. 2020). Finally, chromosome-centric analyses have been found useful in large-scale proteomics research (Archakov et al. 2012). In the postgenomic era, we do have opportunities to describe numerous aspects of human chromosome behavior (Iourov 2019b). Using pathway-based technologies, it becomes possible to have a more precise look at the cellular genome and its behavior (Iourov et al. 2019a). Since the "real" genome basically functions in interphase (Manuelidis 1990; Yurov et al. 2013; Cremer et al. 2020), interphase chromosomes are to be studied in the postgenomic context. Thus, current achievements in genomics made by sequencing and microarraying are able to gain a chromosomal context.

It's All in the Nucleus, Interphase Nucleus

Apart from being useful for uncovering chromosome aberrations, chromosomal bands express the genome organization local to specific chromosome regions: GC content, repeat content, meiotic recombination rate, replication timing, and gene density (Bickmore and Sumner 1989; Costantini et al. 2007; Bernardi 2015). Moreover, chromatin- and DNA-based biophysical (biomechanical) properties of chromosomes (chromosomal loci) are band-specific and shape genome behavior at the level of individual genes or gene sets/clusters (Holmquist 1992; Kosyakova et al. 2009; Watanabe and Maekawa 2013; Daban 2015; Tortora et al. 2020). Indeed, these properties of chromosomes have been systematically observed to determine genome organization in the interphase nucleus (Sadoni et al. 1999; Carvalho et al. 2001; Küpper et al. 2007; Kumar et al. 2020; Tortora et al. 2020). Structurally, chromosome arrangement in interphase is intimately related to chromatin architecture and, thereby, to chromatin remodeling, which is critical for genome activity in a cell (Dixon et al. 2016; Jabbari et al. 2019). More importantly, genome activity (transcription) throughout the cell cycle is modulated by chromosome arrangement and behavior of chromosomal loci in interphase (Cook and Marenduzzo 2018; Cremer and Cremer 2019; Jerković et al. 2020). Additionally, nuclear organization of chromosomes mediates genome safeguarding (DNA damage response, proper chromosome segregation, mitotic checkpoint, etc.), DNA repair and replication, and programmed cell death (Chevret et al. 2000; Rodriguez and Bjerling 2013; Shachar and Misteli 2017; Nagano et al. 2017; Maass et al. 2019; Kumar et al. 2020). In other words, almost all homeostatic processes involving nuclear genomes are connected to chromosome behavior in interphase. Changes in nuclear genome organization have been associated with pathogenic processes in human diseases. These observations have led to proposing diagnostic value of studying spatial genome organization at the chromosomal level (Meaburn 2016; Ouimette et al. 2019). Here, it is to note that spatial arrangement of interphase chromosomes may predispose to chromosomal abnormalities in somatic cells (e.g., translocations), which cause cancers (Maharana et al. 2016; McCord and Balajee 2018; Szczepińska et al. 2019). In summary, CM2C, chromatin studies, and postgenomics have underlain the development of 3D genomics, which aims to understand spatial chromatin/chromosome organization specific to different cell types in health and disease (Meaburn 2016; Shachar and Misteli 2017; Knoch 2019; Jerković et al. 2020; Kumar et al. 2020). Accordingly, two important aspects of studying interphase chromosomes in the genomic context are to be emphasized: (1) chromosome arrangement in interphase is a key to understanding genome behavior, and (2) knowledge about arrangement of interphase chromosome is critical to understand causes and consequences of disease-causing genomic variations.

Genomic Variations: Mind the Chromosome

CM2C continuously generate data on chromosomal abnormalities presented by cohort case-control studies or case reports. Currently, balanced and unbalanced chromosomal rearrangements (including CNVs) represent the commonest type of genomic variations associated with morbid conditions (Gersen and Keagle 2005; Feuk et al. 2006; Iourov et al. 2006a, 2008b, 2019b; Liehr 2013; Gonzales et al. 2016; Ho et al. 2020). Cancer is associated with (somatic) chromosomal aberrations and instability (Heng et al. 2011; Liehr 2017; Christine et al. 2018; Hnisz et al. 2018; Ye et al. 2019). Surprisingly, deserved attention is not currently paid to this fact in human genetics and genomics (Crellin et al. 2019; Whitley et al. 2020). Still, the contribution of chromosomal or subchromosomal (genomic) rearrangements to human morbidity is to be kept in mind.

During the last decades, somatic genome variations manifesting as somatic chromosomal mosaicism have become a major focus of biomedical research. Interestingly, this type of intercellular genomic variations is relevant to a wide spectrum of diseases and morbid conditions (Yurov et al. 2001, 2018b; Iourov et al. 2006a, 2006b, 2008a, 2019c; Vorsanova et al. 2010a, 2010c; Heng et al. 2011). Alternatively, chromosomal mosaicism and instability are mechanisms of natural genomic variation in cellular populations (Yurov et al. 2005, 2007; Iourov et al. 2009b). However, it is generally accepted that clinical populations (intellectual disability and congenital malformations) and fetal specimens exhibit high rates of chromosomal mosaicism (Gersen and Keagle 2005; Yurov et al. 2007; Iourov et al. 2008a, 2019c; Vorsanova et al. 2010c). Chromosomal mosaicism confined to the brain may cause neuropsychiatric diseases (schizophrenia, autism, epilepsy) (Yurov et al. 2001, 2008, 2016, 2018a; Vorsanova et al. 2007). Furthermore, neurodegeneration is mediated by aneuploidy (gains/losses of chromosomes) and chromosome instability confined to degenerating brain areas (Iourov et al. 2009a, 2011; Yurov et al. 2011, 2014, 2019). It appears that behavioral changes/problems are able to be associated with dynamic nature of somatic (chromosomal) mosaicism (Vorsanova et al. 2018). Finally, aging is associated with accumulation of somatic chromosomal mutations (aneuploidy/chromosome losses) as previously mentioned (Yurov et al. 2009b, 2010, 2014). Certainly, the occurrence of somatic mutations is the result of genetic-environmental interactions, which may either generate or inhibit accumulation of somatic chromosomal mutations (Iourov et al. 2013). Fortunately, there are available molecular cytogenetic approaches to analyze chromosomal variations in the environmental context (Hovhannisyan 2010; Iurov et al. 2011). Somatic chromosomal mosaicism requires specific methods of the diagnosis. CM2C research has demonstrated that molecular cytogenetic/cytogenomic monitoring and analysis of postmitotic tissues are required for proper surveying of somatic chromosomal mosaicism (Vorsanova et al. 2010b, 2010c; Liehr 2017; Iourov et al. 2019c). For establishing causes and consequences of chromosome abnormalities (+somatic chromosomal mosaicism and instability), systems biology or bioinformatics analyses are to be applied. These methods may be used to modulate functional consequences of chromosome/genomic imbalances at epigenome, proteome, and metabolome levels or, in other words, unravel disease mechanisms (Iourov et al. 2014a, 2019a, 2019b; Vorsanova et al. 2017; Yurov et al. 2017; Zelenova et al. 2019). For example, in silico molecular cytogenetic technologies appreciably increase diagnostic outcomes of CM2C studies of children with neurodevelopmental diseases (Iourov et al. 2016). In some cases, these may provide unprecedented correlations between phenotypes and molecular karyotypes (i.e., neuropsychological genotype/phenotype correlations) (Iourov et al. 2018). In addition, systems biology analysis of genomic variation allows evaluations of causative alterations to molecular pathways, which are involved in generation of chromosomal aberrations and instability (Iourov et al. 2015a). Thus, according to CM2C studies, effective analysis of the genome requires three technological blocks: visualization (banding cytogenetics + metaphase/interphase FISH; single-cell analyses), whole genome scanning (whole genome microarray and/or sequencing), and bioinformatics (Iourov et al. 2012, 2014b; Vorsanova et al. 2019). In total, basic and diagnostic CM2C analyses seem to become more sophisticated than previously recognized.

Probably the most exciting outcome of the interaction of systems biology and CM2C is the development of therapeutic interventions in diseases resulting from chromosome imbalances and instability, which have been condemned to be incurable (Yurov et al. 2009a; Iourov et al. 2015b; Iourov 2016, 2019a). To succeed regularly, there is a need for a systems biology analysis complemented by data obtained by chromosome-oriented studies offering a chromosome-centric look at the genome behavior. These data should encompass variome (the whole set of genome variations specific for an individual or a disease), methylome, chromatin remodeling and organization, and chromosomal arrangement in the nucleus (Iourov et al. 2012, 2014a, 2019a, 2019b; Yurov et al. 2017; Christine et al. 2018; Knoch 2019; Zelenova et al. 2019; Heng 2020). In our opinion, such processing of systems biology data is better to show using linear algebra. Empirical or theoretical data about effects (interrelations) of genomic variations (mosaic/non-mosaic sequence variations/CNV/chromosome abnormalities and chromosomal/genomic instability) on chromatin behavior and chromosomal nuclear organization may be used for constructing a matrix. The systems biology analysis of this matrix (evaluation of these effects in

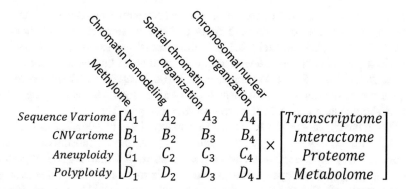

$$\begin{array}{l} Sequence\ Variome \\ CNVariome \\ Aneuploidy \\ Polyploidy \end{array} \begin{bmatrix} A_1 & A_2 & A_3 & A_4 \\ B_1 & B_2 & B_3 & B_4 \\ C_1 & C_2 & C_3 & C_4 \\ D_1 & D_2 & D_3 & D_4 \end{bmatrix} \times \begin{bmatrix} Transcriptome \\ Interactome \\ Proteome \\ Metabolome \end{bmatrix}$$

Fig. 9.1 Basic principle of the systems genome analysis complemented by the chromosome-centric look is schematically shown by matrix multiplication: "(3D) genome-chromosome" matrix, where interrelations between genomic variability (sequence variome, CNVariome, numerical chromosome imbalances) and epigenetic variations + spatial genome organization are designated by letters A_1-A_4, B_1-B_4, C_1-C_4, and D_1-D_4, is "multiplied" by the systems biology row matrix. The idea is that all the data on genome variability and epigenetic variation/spatial genome organization should be considered in the light of systems biology (transcriptome, interactome, proteome, and metabolome)

transcriptome, interactome, proteome, and metabolome context; for more details, see Iourov et al. [2012, 2014a], Yurov et al. [2017]) would be multiplication by row matrix. The operation is shown in Fig. 9.1. The ability to fully interpret CM2C data is required to understand the blueprint of life.

Cytogeneticists: The Superheroes of Translational Biomedicine

There is a perception that cytogenetics is in crisis (Liehr 2013, 2019; Hochstenbach et al. 2017, 2019). "The alarm bell has started to ring" because of appreciable fading competency of European cytogenetic diagnostic laboratories (Hochstenbach et al. 2017). Our experience indicates that "the alarm bell" might ring a decade earlier. Optimistically, it takes 2–5 years for a researcher to become a more or less autonomous in banding cytogenetic analysis. Certainly, theoretical issues in cytogenetics are less time-consuming, but streamlining this knowledge in a biomedical context requires time. Relentless pursuit of simplicity has negatively impacted cytogenetics as a science and as a diagnostic discipline. Thus, it has become less popular among biomedical students due to the complexity of the microscopic analysis and unjustified ignorance of CM2C data by researchers' community from other areas of genomics/medical genetics (Vorsanova et al. 2008). Further ignorance produced by scientifically unsupported simplifications resulted in leaving aside numerous aspects of chromosomes such as constitutive heterochromatin, chromosomal heterogeneity at supramolecular and microscopic levels, and low-level chromosomal mosaicism.

However, heterochromatic regions of human chromosomes represent an intriguing and important matter of genome biology and chromosome research (Vorsanova et al. 2007, 2008, 2010a; Liehr 2013). Chromosomal heterogeneity and low-level mosaicism are phenomena relevant to numerous areas of bioscience (Iourov et al. 2008a, 2010, 2014b, 2019c; Vorsanova et al. 2010c; Heng 2020). Therefore, there is a need for more effective and student friendly educational efforts in cytogenetics, which should encompass CM2C in the widest sense possible.

Facing the challenges of CM2C, we have to envision a new kind of superhero in an attempt to portray today's cytogeneticist. In a postgenomic (postmodern) world, cytogeneticist performing basic and diagnostic research should able to (i) perform banding cytogenetic analysis (classics), (ii) handle microarray data (at least at the elementary level), (iii) understand the meaning of chromosomal bands and chromosome variations in the genomic context and in the context of the spatial arrangement in interphase nuclei (Bickmore and Sumner 1989; Sadoni et al. 1999; Iourov et al. 2006a; Kosyakova et al. 2009; Vorsanova et al. 2010b; Watanabe and Maekawa 2013; Yurov et al. 2013; Bernardi 2015; Daban 2015; Cremer et al. 2020), (iv) process CM2C data using systems biology methodology (Iourov et al. 2014a, 2019b; Vorsanova et al. 2017; Yurov et al. 2017; Iourov 2019a; Zelenova et al. 2019), (v) apply molecular cytogenetic techniques (e.g., FISH) to uncover chromosomal mosaicism and instability (Vorsanova et al. 2010c, 2019; Iourov et al. 2012), (vi) interpret/correlate data on non-mosaic genomic variations and chromosomal mosaicism/instability (Iourov et al. 2015a, 2019a; Heng 2020), and (vii) envisage the applications of single-cell genomic technologies (Iourov et al. 2012; McClelland 2019). In fact, modern cytogeneticists are those who are able to ensure translational nature of current biomedicine, i.e., connecting human genetics and genomics, molecular and cell biology, evolution, oncology, and systems biology. That is a bit too much, isn't it? Some competencies are likely to be distributed among the personnel. Figure 9.2 shows the "quest" of such a "superhero" a.k.a. true cytogeneticist.

Conclusion: Learn, Learn, and Learn

Taking into account the way cytogenetics goes (Vorsanova et al. 2008; Hochstenbach et al. 2017, 2019; Liehr 2019) and the requirements cytogeneticists have to meet, a question arises: What is to be done? The question becomes even more crucial when current education programs in genomics are considered. For instance, the latest descriptions of education programs in genomics and genomic medicine for biomedical students, physicians, or the public do not have any information about chromosomes (Korf et al. 2014; Crellin et al. 2019; McClaren et al. 2020; Whitley et al. 2020). The word "chromosome" is absent. Hence, we conclude that education is an urgent and significant issue in current cytogenetics.

The expected detriment of ignorance of CM2C studies and cytogenetics as an important area of biomedical research has already led to negative outcomes. Basic

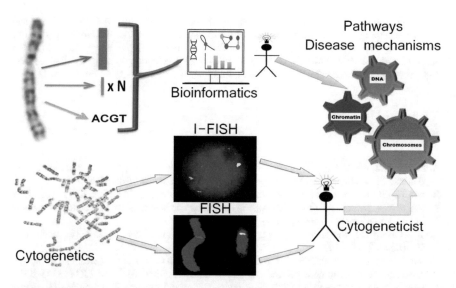

Fig. 9.2 The quest of cytogeneticists in a postmodern world. A cytogeneticist has to acquire and analyze data using cytogenetic single-cell analyses: chromosome banding (e.g., metaphase plate in the lower left) and FISH-based techniques (i.e., metaphase FISH and I-FISH). These data require an appreciable experience in the field of cytogenetics; otherwise, the data are useless for further system biology analyses to discover molecular pathways to diseases (disease mechanisms). In addition, CM2C is also the study of microarray and, in some instances, sequencing data (schematically shown in the upper left). These data require bioinformatics' approaches to acquire and interpret the data. Finally, cytogeneticists (a.k.a. molecular cytogeneticists or "cytogenomicists") have to integrate these two massive data sets for each individual sample to identify causes and consequences of genomic variations at the chromosomal level

chromosome research losses the genomic context and vice versa. Cytogenetic diagnostics is shown to be affected by fading competency. A need to recover from these impacts appears to exist. We suggest that a promotion of chromosome-centric look at the genome or genomic data may help. To succeed, the aforementioned problems are likely to be solved, when considered in four dimensions: theoretical, empirical, diagnostic, and educational. Theoretical solutions for current cytogenetics are based on integration of CM2C data processed by systems biology protocols. The experimental solutions are rooted from putting CM2C data in the genomic context and vice versa putting genomic data (data on genes) in the chromosomal context. Improvements in molecular diagnosis of causative genome variations have to come from developments in translational medicine, which provided effective co-application of visualization, genome scanning, and bioinformatics techniques. Out of the chromosomal context, genomic data are poorly applicable for unraveling disease mechanisms and developing therapeutic strategies. Finally, theoretical, empirical, and diagnostic solutions are not guaranteed without the educational one, which should not disintegrate human genetics/genomics into specialized and disconnected parts attributed to either genome biology or chromosome research.

Acknowledgments Professors SG Vorsanova and IY Iourov are partially supported by RFBR and CITMA according to the research project No. 18-515-34005. Prof. IY Iourov's lab is supported by the Government Assignment of the Russian Ministry of Science and Higher Education, Assignment no. AAAA-A19-119040490101-6. Prof. SG Vorsanova's lab is supported by the Government Assignment of the Russian Ministry of Health, Assignment no. AAAA-A18-118051590122-7.

References

Archakov A, Zgoda V, Kopylov A et al (2012) Chromosome-centric approach to overcoming bottlenecks in the Human Proteome Project. Expert Rev Proteomics 9(6):667–676

Bernardi G (2015) Chromosome architecture and genome organization. PLoS One 10(11):e0143739

Bickmore WA, Sumner AT (1989) Mammalian chromosome banding – an expression of genome organization. Trends Genet 5(5):144–148

Bridges CB (1916) Non-disjunction as proof of the chromosome theory of heredity (concluded). Genetics 1(2):107–163

Carvalho C, Pereira HM, Ferreira J et al (2001) Chromosomal G-dark bands determine the spatial organization of centromeric heterochromatin in the nucleus. Mol Biol Cell 12(11):3563–3572

Chevret E, Volpi EV, Sheer D (2000) Mini review: form and function in the human interphase chromosome. Cytogenet Cell Genet 90(1–2):13–21

Christine JY, Regan S, Liu G et al (2018) Understanding aneuploidy in cancer through the lens of system inheritance, fuzzy inheritance and emergence of new genome systems. Mol Cytogenet 11:31

Cook PR, Marenduzzo D (2018) Transcription-driven genome organization: a model for chromosome structure and the regulation of gene expression tested through simulations. Nucleic Acids Res 46(19):9895–9906

Costantini M, Clay O, Federico C et al (2007) Human chromosomal bands: nested structure, high-definition map and molecular basis. Chromosoma 116(1):29–40

Crellin E, McClaren B, Nisselle A et al (2019) Preparing medical specialists to practice genomic medicine: education an essential part of a broader strategy. Front Genet 10:789

Cremer M, Cremer T (2019) Nuclear compartmentalization, dynamics, and function of regulatory DNA sequences. Genes Chromosomes Cancer 58(7):427–436

Cremer T, Cremer M, Hübner B et al (2020) The interchromatin compartment participates in the structural and functional organization of the cell nucleus. BioEssays 42(2):e1900132

Daban JR (2015) Stacked thin layers of metaphase chromatin explain the geometry of chromosome rearrangements and banding. Sci Rep 5:14891

Dixon JR, Gorkin DU, Ren B (2016) Chromatin domains: the unit of chromosome organization. Mol Cell 62(5):668–680

Feuk L, Marshall CR, Wintle RF et al (2006) Structural variants: changing the landscape of chromosomes and design of disease studies. Hum Mol Genet 15(Sup.1):R57–R66

Gersen SL, Keagle MB (2005) The principles of clinical cytogenetics, 2nd edn. Humana Press, Totowa

Gonzales PR, Carroll AJ, Korf BR (2016) Overview of clinical cytogenetics. Curr Protoc Hum Genet 89:8.1.1–8.1.13

Heng HH (2020) New data collection priority: focusing on genome-based bioinformation. Res Result Biomed 6(1):5–8

Heng HH, Liu G, Stevens JB et al (2011) Decoding the genome beyond sequencing: the new phase of genomic research. Genomics 98(4):242–252

Hnisz D, Schuijers J, Li CH et al (2018) Regulation and dysregulation of chromosome structure in cancer. Annu Rev Cancer Biol 2:21–40

Ho SS, Urban AE, Mills RE (2020) Structural variation in the sequencing era. Nat Rev Genet 21(3):171–189

Hochstenbach R, Slunga-Tallberg A, Devlin C et al (2017) Fading competency of cytogenetic diagnostic laboratories: the alarm bell has started to ring. Eur J Hum Genet 25(3):273–274

Hochstenbach R, van Binsbergen E, Schuring-Blom H et al (2019) A survey of undetected, clinically relevant chromosome abnormalities when replacing postnatal karyotyping by Whole Genome Sequencing. Eur J Med Genet 62(9):103543

Holmquist GP (1992) Chromosome bands, their chromatin flavors, and their functional features. Am J Hum Genet 51(1):17–37

Hovhannisyan GG (2010) Fluorescence in situ hybridization in combination with the comet assay and micronucleus test in genetic toxicology. Mol Cytogenet 3:17

Iourov IY (2016) Post genomics: towards a personalized approach to chromosome abnormalities. J Down Syndr Chromosom Abnorm 2(1):2:e104

Iourov IY (2019a) Cytogenomic bioinformatics: practical issues. Curr Bioinformatics 14(5):372–373

Iourov IY (2019b) Cytopostgenomics: what is it and how does it work? Curr Genomics 20(2):77–78

Iourov IY, Vorsanova SG, Yurov YB (2006a) Chromosomal variation in mammalian neuronal cells: known facts and attractive hypotheses. Int Rev Cytol 249:143–191

Iourov IY, Vorsanova SG, Yurov YB (2006b) Intercellular genomic (chromosomal) variations resulting in somatic mosaicism: mechanisms and consequences. Curr Genomics 7(7):435–446

Iourov IY, Vorsanova SG, Yurov YB (2008a) Chromosomal mosaicism goes global. Mol Cytogenet 1:26

Iourov IY, Vorsanova SG, Yurov YB (2008b) Molecular cytogenetics and cytogenomics of brain diseases. Curr Genomics 9(7):452–465

Iourov IY, Vorsanova SG, Liehr T et al (2009a) Increased chromosome instability dramatically disrupts neural genome integrity and mediates cerebellar degeneration in the ataxia-telangiectasia brain. Hum Mol Genet 18(14):2656–2669

Iourov IY, Vorsanova SG, Liehr T et al (2009b) Aneuploidy in the normal, Alzheimer's disease and ataxia-telangiectasia brain: differential expression and pathological meaning. Neurobiol Dis 34(2):212–220

Iourov IY, Vorsanova SG, Yurov YB (2010) Somatic genome variations in health and disease. Curr Genomics 11(6):387–396

Iourov IY, Vorsanova SG, Yurov YB (2011) Genomic landscape of the Alzheimer's disease brain: chromosome instability – aneuploidy, but not tetraploidy – mediates neurodegeneration. Neurodegener Dis 8(1–2):35–37

Iourov IY, Vorsanova SG, Yurov YB (2012) Single cell genomics of the brain: focus on neuronal diversity and neuropsychiatric diseases. Curr Genomics 13(6):477–488

Iourov IY, Vorsanova SG, Yurov YB (2013) Somatic cell genomics of brain disorders: a new opportunity to clarify genetic-environmental interactions. Cytogenet Genome Res 139(3):181–188

Iourov IY, Vorsanova SG, Yurov YB (2014a) In silico molecular cytogenetics: a bioinformatic approach to prioritization of candidate genes and copy number variations for basic and clinical genome research. Mol Cytogenet 7(1):98

Iourov IY, Vorsanova SG, Liehr T et al (2014b) Mosaike im Gehirn des Menschen. Diagnostische Relevanz in der Zukunft? Med Genet 26:342–345

Iourov IY, Vorsanova SG, Zelenova MA et al (2015a) Genomic copy number variation affecting genes involved in the cell cycle pathway: implications for somatic mosaicism. Int J Genomics 2015:757680

Iourov IY, Vorsanova SG, Voinova VY et al (2015b) 3p22.1p21.31 microdeletion identifies CCK as Asperger syndrome candidate gene and shows the way for therapeutic strategies in chromosome imbalances. Mol Cytogenet 8:82

Iourov IY, Vorsanova SG, Korostelev SA et al (2016) Structural variations of the genome in autistic spectrum disorders with intellectual disability. Zh Nevrol Psikhiatr Im S S Korsakova 116(7):50–54

Iourov IY, Zelenova MA, Vorsanova SG et al (2018) 4q21.2q21.3 duplication: molecular and neuropsychological aspects. Curr Genomics 19(3):173–178

Iourov IY, Vorsanova SG, Yurov YB (2019a) Pathway-based classification of genetic diseases. Mol Cytogenet 12:4

Iourov IY, Vorsanova SG, Yurov YB (2019b) The variome concept: focus on CNVariome. Mol Cytogenet 12:52

Iourov IY, Vorsanova SG, Yurov YB et al (2019c) Ontogenetic and pathogenetic views on somatic chromosomal mosaicism. Genes (Basel) 10(5):E379

Iurov II, Vorsanova SG, Solov'ev IV et al (2011) Original molecular cytogenetic approach to determining spontaneous chromosomal mutations in the interphase cells to evaluate the mutagenic activity of environmental factors. Gig Sanit 5:90–94

Jabbari K, Wirtz J, Rauscher M et al (2019) A common genomic code for chromatin architecture and recombination landscape. PLoS One 14(3):e0213278

Jerković I, Szabo Q, Bantignies F et al (2020) Higher-order chromosomal structures mediate genome function. J Mol Biol 432(3):676–681

Knoch TA (2019) Simulation of different three-dimensional polymer models of interphase chromosomes compared to experiments – an evaluation and review framework of the 3D genome organization. Semin Cell Dev Biol 90:19–42

Korenberg JR, Rykowski MC (1988) Human genome organization: Alu, lines, and the molecular structure of metaphase chromosome bands. Cell 53(3):391–400

Korf BR, Berry AB, Limson M et al (2014) Framework for development of physician competencies in genomic medicine: report of the Competencies Working Group of the Inter-Society Coordinating Committee for Physician Education in Genomics. Genet Med 16(11):804–809

Kosyakova N, Weise A, Mrasek K et al (2009) The hierarchically organized splitting of chromosomal bands for all human chromosomes. Mol Cytogenet 2:4

Kumar Y, Sengupta D, Bickmore W (2020) Recent advances in the spatial organization of the mammalian genome. J Biosci 45:18

Küpper K, Kölbl A, Biener D (2007) Radial chromatin positioning is shaped by local gene density, not by gene expression. Chromosoma 116(3):285–306

Liehr T (2013) Benign and pathological chromosomal imbalances: microscopic and submicroscopic copy number variations (CNVs) in genetics and counseling. Academic Press

Liehr T (2017) Fluorescence in situ hybridization (FISH) – application guide. Springer, Berlin/ Heidelberg

Liehr T (2019) From human cytogenetics to human chromosomics. Int J Mol Sci 20(4):E826

Maass PG, Barutcu AR, Rinn JL (2019) Interchromosomal interactions: a genomic love story of kissing chromosomes. J Cell Biol 218(1):27–38

Maharana S, Iyer KV, Jain N et al (2016) Chromosome intermingling – the physical basis of chromosome organization in differentiated cells. Nucleic Acids Res 44(11):5148–5160

Manuelidis L (1990) A view of interphase chromosomes. Science 250(4987):1533–1540

McClaren BJ, Crellin E, Janinski M et al (2020) Preparing medical specialists for genomic medicine: continuing education should include opportunities for experiential learning. Front Genet 11:151

McClelland SE (2019) Single-cell approaches to understand genome organization throughout the cell cycle. Essays Biochem 63(2):209–216

McCord RP, Balajee A (2018) 3D genome organization influences the chromosome translocation pattern. Adv Exp Med Biol 1044:113–133

Meaburn KJ (2016) Spatial genome organization and its emerging role as a potential diagnosis tool. Front Genet 7:134

Nagano T, Lubling Y, Várnai C et al (2017) Cell-cycle dynamics of chromosomal organization at single-cell resolution. Nature 547(7661):61–67

Ouimette JF, Rougeulle C, Veitia RA (2019) Three-dimensional genome architecture in health and disease. Clin Genet 95(2):189–198

Page RD, Holmes EC (2009) Molecular evolution: a phylogenetic approach. Wiley

Rodriguez A, Bjerling P (2013) The links between chromatin spatial organization and biological function. Biochem Soc Trans 41(6):1634–1639

Sadoni N, Langer S, Fauth C et al (1999) Nuclear organization of mammalian genomes. Polar chromosome territories build up functionally distinct higher order compartments. J Cell Biol 146(6):1211–1226

Savory K, Manivannan S, Zaben M et al (2020) Impact of copy number variation on human neurocognitive deficits and congenital heart defects: a systematic review. Neurosci Biobehav Rev 108:83–93

Shachar S, Misteli T (2017) Causes and consequences of nuclear gene positioning. J Cell Sci 130(9):1501–1508

Szczepińska T, Rusek AM, Plewczynski D (2019) Intermingling of chromosome territories. Genes Chromosomes Cancer 58(7):500–506

Tortora MM, Salari H, Jost D (2020) Chromosome dynamics during interphase: a biophysical perspective. Curr Opin Genet Dev 61:37–43

Umbreit NT, Zhang CZ, Lynch LD et al (2020) Mechanisms generating cancer genome complexity from a single cell division error. Science 368(6488):eaba0712

Vorsanova SG, Yurov IY, Demidova IA et al (2007) Variability in the heterochromatin regions of the chromosomes and chromosomal anomalies in children with autism: identification of genetic markers of autistic spectrum disorders. Neurosci Behav Physiol 37(6):553–558

Vorsanova SG, Yurov IY, Soloviev IV et al (2008) Geterokhromatinovye raiony khromosom cheloveka: klinikobiologicheskie aspekty (Heterochromatic regions of human chromosomes: clinico-biological aspects). Medpraktika, Moscow

Vorsanova SG, Voinova VY, Yurov IY et al (2010a) Cytogenetic, molecular-cytogenetic, and clinical-genealogical studies of the mothers of children with autism: a search for familial genetic markers for autistic disorders. Neurosci Behav Physiol 40(7):745–756

Vorsanova SG, Yurov YB, Iourov IY (2010b) Human interphase chromosomes: a review of available molecular cytogenetic technologies. Mol Cytogenet 3:1

Vorsanova SG, Yurov YB, Soloviev IV et al (2010c) Molecular cytogenetic diagnosis and somatic genome variations. Curr Genomics 11(6):440–446

Vorsanova SG, Yurov YB, Iourov IY (2017) Neurogenomic pathway of autism spectrum disorders: linking germline and somatic mutations to genetic-environmental interactions. Curr Bioinforma 12(1):19–26

Vorsanova SG, Zelenova MA, Yurov YB et al (2018) Behavioral variability and somatic mosaicism: a cytogenomic hypothesis. Curr Genomics 19(3):158–162

Vorsanova SG, Yurov YB, Soloviev IV et al (2019) FISH-based analysis of mosaic aneuploidy and chromosome instability for investigating molecular and cellular mechanisms of disease. OBM Genetics 3(1):9

Watanabe Y, Maekawa M (2013) R/G-band boundaries: genomic instability and human disease. Clin Chim Acta 419:108–112

Whitley KV, Tueller JA, Weber KS (2020) Genomics education in the era of personal genomics: academic, professional, and public considerations. Int J Mol Sci 21(3):E768

Ye CJ, Stilgenbauer L, Moy A et al (2019) What is karyotype coding and why is genomic topology important for cancer and evolution? Front Genet 10:1082

Yurov YB, Vostrikov VM, Vorsanova SG et al (2001) Multicolor fluorescent in situ hybridization on post-mortem brain in schizophrenia as an approach for identification of low-level chromosomal aneuploidy in neuropsychiatric diseases. Brain Dev 23(Sup.1):S186–S190

Yurov YB, Iourov IY, Monakhov VV et al (2005) The variation of aneuploidy frequency in the developing and adult human brain revealed by an interphase FISH study. J Histochem Cytochem 53(3):385–390

Yurov YB, Iourov IY, Vorsanova SG et al (2007) Aneuploidy and confined chromosomal mosaicism in the developing human brain. PLoS One 2(6):e558

Yurov YB, Iourov IY, Vorsanova SG et al (2008) The schizophrenia brain exhibits low-level aneuploidy involving chromosome 1. Schizophr Res 98(1–3):139–147

Yurov YB, Iourov IY, Vorsanova SG (2009a) Neurodegeneration mediated by chromosome instability suggests changes in strategy for therapy development in ataxia-telangiectasia. Med Hypotheses 73(6):1075–1076

Yurov YB, Vorsanova SG, Iourov IY (2009b) GIN'n'CIN hypothesis of brain aging: deciphering the role of somatic genetic instabilities and neural aneuploidy during ontogeny. Mol Cytogenet 2:23

Yurov YB, Vorsanova SG, Iourov IY (2010) Ontogenetic variation of the human genome. Curr Genomics 11(6):420–425

Yurov YB, Vorsanova SG, Iourov IY (2011) The DNA replication stress hypothesis of Alzheimer's disease. ScientificWorldJournal 11:2602–2612

Yurov YB, Vorsanova SG, Iourov IY (2013) Human interphase chromosomes – biomedical aspects. Springer, New York/Heidelberg/Dordrecht/London

Yurov YB, Vorsanova SG, Liehr T et al (2014) X chromosome aneuploidy in the Alzheimer's disease brain. Mol Cytogenet 7(1):20

Yurov YB, Vorsanova SG, Demidova IA et al (2016) Genomic instability in the brain: chromosomal mosaicism in schizophrenia. Zh Nevrol Psikhiatr Im S S Korsakova 116(11):86–91

Yurov YB, Iourov IY, Vorsanova SG (2017) Network-based classification of molecular cytogenetic data. Curr Bioinforma 12(1):27–33

Yurov YB, Vorsanova SG, Demidova IA et al (2018a) Mosaic brain aneuploidy in mental illnesses: an association of low-level post-zygotic aneuploidy with schizophrenia and comorbid psychiatric disorders. Curr Genomics 19(3):163–172

Yurov YB, Vorsanova SG, Iourov IY (2018b) Human molecular neurocytogenetics. Curr Genet Med Rep 6(4):155–164

Yurov YB, Vorsanova SG, Iourov IY (2019) Chromosome instability in the neurodegenerating brain. Front Genet 10:892

Zelenova MA, Yurov YB, Vorsanova SG et al (2019) Laundering CNV data for candidate process prioritization in brain disorders. Mol Cytogenet 12:54

Index

© Springer Nature Switzerland AG 2020
I. Iourov et al. (eds.), *Human Interphase Chromosomes*,
https://doi.org/10.1007/978-3-030-62532-0

Printed in the United States
by Baker & Taylor Publisher Services